THE JOOM DESTINY

JUST ON ORDER MAKING

*How 3D Printing
Will Revolutionize Your World*

Dr. Future

Produced by:

FriesenPress

Suite 300 – 852 Fort Street

Victoria, BC, Canada V8W 1H8

www.friesenpress.com

Distributed to the trade by The Ingram Book Company

Contents

Get Ready for the **New** Industrial Revolution

JOOM Represents a Fundamental Paradigm Shift

JOOM Will Change Every Aspect of Human Society

JOOM Will Change Your Life Forever

JOOM is Happening Right Now

Will You Be Ready?

About the Author

Figure 1: The author just before his now historic speech at the Naval War College in Newport, RI. Source: Tinari

Dr. Future (aka Dr. Paul Tinari) has been called a visionary and "Renaissance Man" for the 21st century. When asked by a U.S. Senator what his research would do to enhance the national security of the nation, he replied: "Nothing, except to help make it worth defending." He founded the Pacific Institute for Advanced Study in 1990, the world's first truly globally networked R&D organization. His curiosity and research interests have spanned many domains including alternate energy, architecture, art, archaeology, biology, engineering, epidemiology, environmental science, fluid dynamics, future studies, green buildings, kinesiology, mathematics, physics and zoology. The common thread that runs through all of his interests is an intense desire to improve the world and to help people lead better, safer, healthier and more abundant lives. Among his numerous accomplishments include the organization of the first Canadian Earth Day (http://www.youtube.com/watch?v=cFh7HkGcq0g), consulting with senior administrators at NASA on how new technologies could impact the agency's business model (http://youtu.be/acCc38UcYYA), speaking at the World Future Society Conference on the possible impacts of 3D printing and related technologies (http://www.youtube.com/watch?v=K1KNQN0BYuc), being a finalist in the Canadian astronaut selection process and presenting to the US Navy on how 3D printing and supporting technologies could dramatically reduce the costs of building ships and submarines. In addition to having taught Future Studies at Simon Fraser University, he designs and leads seminars for executives of Fortune 500 corporations in many areas including Creative Thinking, Systems Thinking and on how future technological trends will impact business operations. He currently lives in Port Moody, British Columbia.

Acknowledgement

If I was able to see farther than others, it is because I stood on the shoulders of giants. During the course of the two year project that was the creation of this book, I received a huge amount of help from many highly talented individuals. I would like to thank all my business partners, friends and acquaintances for their tireless help, their encouragement, and for their ability to keep me focused on all of the endless details that had to be attended to. From the depths of my heart I thank each of you for all the work that you did editing the various chapters of the manuscript and for weeding out the numerous errors that were found there. Without each of your contributions, this book could never have been completed. Thank You.

Chapter 1
Introduction

The tall, thin man stood confidently at the front of the lecture hall located in the *Naval War College* in Newport, RI. He was dressed in a plain dark suit that was a half a size too large for his slender frame. He spoke in a strong, assertive voice that carefully enunciated each syllable, while his eyes continuously scanned the serious, experience laden faces that were listening to him intently. The audience consisted only of senior officers, of rank navy captain and above, centered on James Hogg, a retired four star admiral who was the Director of the *CNO Strategic Studies Group*. Admiral Hogg was acting as chair for this particular meeting. The man they were all listening to was not in the navy, nor was he even in the US military. Rather, he was a civilian of French Canadian decent, who was the Director of the Vancouver based Pacific Institute for Advanced Study. He had been invited to address this group in an attempt to generate some radically new ideas that could be used to develop future naval capabilities.

Many of the officers in the room had commanded ships that had been part of aircraft carrier battle groups. They were fully aware that before it could set out for sea, each battle group had to be loaded with tens of millions of spare parts that would allow all of the components of the flotilla to continue functioning seamlessly in the event of a failure. If there were multiple failures involving a particular part during a lengthy deployment, the entire viability of a critical system could be compromised, perhaps even threatening the mission. Admiral Hogg knew this was an unacceptable situation and he had made it clear that the Navy had made it a high priority to find a solution. But what could be done? Ships had carried all of the spare parts they could possibly need for their sea deployments since the time of the Greek Triremes.

The sole civilian in the room explained that traditionally, the manufacturing of individual parts was carried out by starting with a solid block of raw material and then removing the excess until the final part was revealed. Alternately, the raw material would be melted and then poured into a mould that had been laboriously prepared by a variety of methods. After the material had solidified, out would pop a part with the exact dimensions desired. The whole industrial revolution was based on manufacturing parts for machines based on these two approaches.

But, noted the speaker, rapid development of computer technologies in the 1990s had permitted the parallel development of software capable of designing and visualizing objects in three dimensions. This in turn permitted a new type of manufacturing where the software would first divide a virtual 3D part into a multitude of thin, parallel, slices. The dimensions for each of these slices would be transmitted to a machine equipped with a nozzle that would deposit a corresponding thin layer of material exactly matching the shape in computer memory. The entire part would be built up additively, layer-by-layer until the entire part was produced.

Now looking directly at Admiral Hogg, the speaker concluded that they had the solution to Navy's spare parts problem staring right at them. Additive manufacturing, as it was then known, could be used to construct parts at sea, just as they were needed. Starting with a raw material consisting of sintered (powdered) metal such as titanium or stainless steel, a 3D printer using a high power laser or an electron beam could be used to make any part desired. The machines themselves could be made as large as necessary, so that eventually, turbine engines, aircraft and even entire ships could be built on demand. In fact, there appeared to be no limit to a ultimate capabilities of the new technology.

The impacts of this new technology are what this book is all about. But before this technology and the profound impacts that it will have on society can be examined in depth, it is necessary to take a brief historical tour to see how we got to where we are today, so that a better understanding can be had of where we will possibly go tomorrow.

Historical Preface – The Evolution of Manufacturing

As big-brained, tool-using creatures, humans gained a significant advantage over physically much larger and more powerful creatures. It was the ancients who discovered that primitive tools could be used to create machines that would make life more pleasant by reducing the amount of human labour required to accomplish a specific task. Thousands of years ago it was discovered that no matter how complex the machine, it was essentially made of at most five simpler machines, namely the lever, wheel, axle, pulley and the wedge/screw.

The first machines were constructed from commonly available natural materials such as stone, wood, skin, ivory or bone. With the discovery of metal, it became possible to build stronger and far more efficient machines that further reduced the requirement for human input. With the invention of the sail, for the first time primitive boats had easier access to distant shores without the need for human muscular effort, leading the establishment of international trade. Combining the idea of the sail with the wheel led to the invention of windmills. Bringing together the wheel, axle, lever, screw and the wedge led to the development of the water wheel.

Mass production could be defined as: "*A system of industrial production involving the manufacture of a product or part in large quantities at comparatively low unit cost.*" The advantages of mass production and standardization were perceived early. The fact that beads and other forms of bodily ornaments found in Neanderthal graves were observed to be in standardized shapes and sizes gave evidence for the existence of some form of primitive assemble lines as long ago as 30,000 years. The first attempts at using machines for mass production as a replacement for human labour were carried out by the Romans late in the Imperial period. To fill the need for more bread to feed the growing populations in the region of Arles, engineers built a series of flour mills down a hillside to exploit waterpower on a massive scale. The sixteen millstones of the Barbegal flour mills were able to produce more than 25 tons of flour per day, enough to feed more than 75,000 citizens.

In addition to grinding grain, the Romans creatively used water wheels to saw logs. A schematic of this technology is shown below.

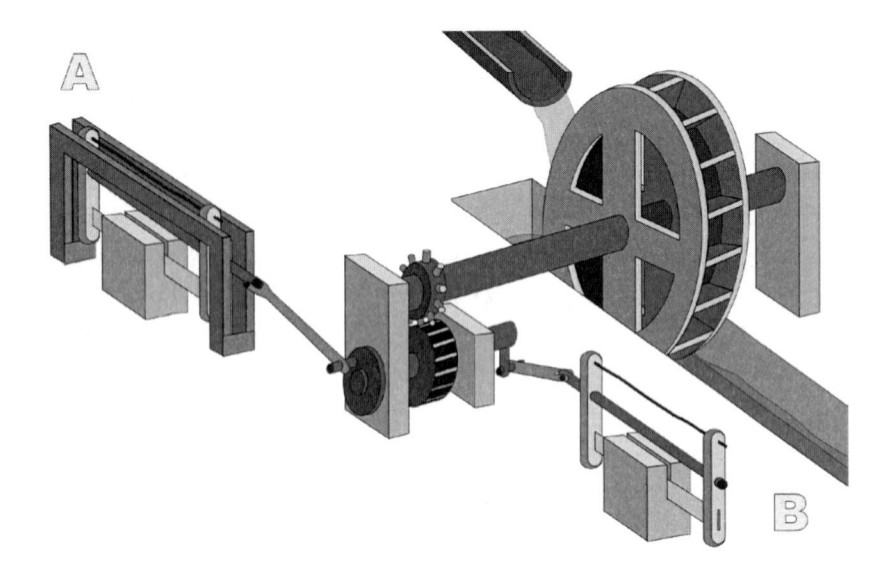

Figure 1-1: Roman water wheel driving two saws. Source: chris

More than 1,500 years ago in China, windmills and water wheels were crushing ore, sawing logs, driving bellows for metals forges, drilling holes for water and gas wells, making paper and drawing fine wire. By the 15th century in Holland, thousands of wind mills dotted the landscape, many driving water pumps to reclaim new lands from the sea.

Notwithstanding the Roman flour mills described above, the idea of exact standardization and mass manufacturing seems to have escaped the ancients. While ancient blacksmiths made thousands of items such as swords to equip entire armies, each piece was a unique artistic creation that had small variations in dimensions making it different from other items made by the same smith in the same forge. However, the benefits of standardization were recognized by the early Middle Ages.

In 789 AD the emperor Charlemagne decided to standardize the hand writing of the scribes by issuing an edict decreeing that only a fixed style of hand writing could be used in all official texts. The goal was to minimize the vagaries of individual handwriting styles to make reading the texts easier. This style endured for centuries, eventually developing into the Gothic script used by the original German printers.

Medieval drawings from the Black Forest region of Germany pictured water powered drop hammers that were used to crush newly extracted ore. The conversion of the rotary motion of the water wheel into the reciprocating up and down movement of the hammers was a significant advance in the history of the mechanization of the means of production.

In about the year 1400, an Italian nobleman hired about 45 scribes to make hand-copies of the various books in his private library. Their maximum output averaged about 100 volumes per year. With the invention of printing in the West in the mid 1400s, the output of a single shop reached more than 300 pages per day, with each copy being an exact duplicate of every other page produced in a given production run. As a result, books which once had been so expensive that only the well-to-do could even dream of owning them, could now became mass consumption items. In addition, the movable-type characters that had made printing possible represented a critical feature needed to implement mass production technology, namely the use of standardized, reusable parts.

Leonardo da Vinci applied his genius to the problem of mass production. Having conceived of a high capacity needle polishing machine, Leonardo wrote in his notebook:

> "Early tomorrow, an. 2, 1496, I shall…proceed to a trial…One hundred times in each hour 400 needles will be finished, making 40,000 in an hour, 480,000 in 12

hours. Suppose we say 4,000 thousands at 5 solidi per thousand gives 20,000 solidi: 1000 lire per working day, and if one works 20 days in the month 60,000 ducats the year."

Figure 1-2: Leonardo da Vinci invented many of the machines that would eventually launch the industrial revolution. This is a model of a sawing machine. Photo Credit: Wang65

Toward the end of the 1500s, a British curate called William Lee invented a knitting machine that could whip out more than 1000 stitches per minute, compared to the 100 or so that the fastest hand craftsman could manage. Being an inventor in the early stages of the Industrial Revolution could often be a hazardous occupation and occasionally, even a fatal one, when workers displaced by the new machines decided to strike back.

In 1596 the city council of Danzig hired a hit-man to strangle the inventor of the ribbon-loom that had put so many of their fellow citizens out of work. John Kay, the inventor of the flying shuttle, had his home wrecked by English textile workers in 1753. Mobs in the early industrial town of Nottingham known as *"Luddites"* cried *"death to the machines"* as they smashed textile-mill steam engines in 1815. In France at around the same period, workers sometimes threw their shoes, called *"sabots,"* into the machinery, in an attempt to destroy it. These were considered to be the first acts of *"sabotage."*

It should be noted in passing that in 1635 an Italian mathematician named Bonaventura Cavalieri observed that any plane could be considered to be made up of an infinite number of parallel lines and that any solid could be constructed from an infinite number of such planes. Without knowing it, Cavalieri had become the first person to imagine how any object could be broken down into any desired number of thin, parallel layers so that it could be constructed with a 3D printer.

To this day, workers continue to attempt to halt the inexorable advance of the machine by symbolic acts of defiance. In 2011, capital cities round the world witnessed the establishment of unsightly "protest camps," mostly consisting of random assortments of high-end alpine tents nestled together with scavenged tarp shelters. Occupied mostly by unemployed or under-employed youth, while the specific aims of the protest camps remained fuzzy, the unifying theme seemed to be people who objected to what they considered to be the obscene concentration of wealth by the banks and multi-national corporations and the feeling that they were being marginalized by the stunning rapidity of global change.

The Elements of Modern Mass Production

Mass production was a synthesis of five different elements:

- **Division of Labour**: Breaking down the production process into separate tasks performed by specially dedicated humans and/or machines that did nothing else

- **Standardization of Parts**: All the individual parts that were mass-produced by machines were built to a specific tolerance, were all interchangeable and could be brought together into more complex assemblies by unskilled workers

- **Precision-Tooling**: Standardization made possible through dies, moulds or machines exactly reproduced according to blueprints that carefully specified every dimension and tolerance

- **The Assembly Line**: A line flow method for moving the work from one worker or machine to the next, evenly and carefully adjusted so that each step could be accomplished within a specific time interval.

- **Mass Demand**: Without mass consumer demand along with mass product transportation and distribution networks, mass production had no logical purpose

The division of labour was the first of the five elements to appear. From the beginning of time, men hunted for game while women gathered berries and then prepared the meat for consumption. After the rise of the agricultural economy, women carded wool and spun it into yarn. Men then took the yarn and wove it into cloth.

In about the year 1700, Christopher Polhem established what was considered to one of the first metal products factories in Stjernsund, Sweden. His innovation was to establish a strict division of labour by training each of his 200 employees in a particular skill and/or teaching them how to operate a particular machine. The result of Polhem's innovation was a dramatic reduction in production costs that resulted in a significant increase in the demand for his products. From the various factories active on the site by 1720, Polhem produced a broad spectrum of useful products including pots and pans, tools, cutlery, screws, nuts & bolts, metal roofing and supplies for the military such as foldable cots. Unfortunately, Polhem was a genius who was too far ahead of his time and sadly, when he died his enterprise died with him.

While Polhem's enterprise disappeared along with him, his ideas fortunately did not. In his timeless epic "*The Wealth of Nations*," Adam Smith observed that ten unskilled men, each working alone, would find it difficult to make one manufactured item each day per man. However, a team of the same ten men working along one factory assembly line with an optimized division of labour could produce tens of thousands of copies of the same item.

In 19th century, the rapidly expanding populations of North American cites required that revolutionary changes had to occur in livestock processing if food supplies were to keep up. Consequently, assembly line methods were applied to every phase of meat processing. One man killed the animal, two lifted and hung the dead animal onto a moving overhead trolley, and another disembowelled the carcass, while another man cleaned it before passing it along to other men for cutting. A new animal entered the first stage of the assembly line every thirty seconds or so.

For hundreds of years, muskets and pistols were individually crafted, expensive works of art. One gun, usually commissioned by a wealthy patron, could take weeks for one highly skilled craftsman to complete. The failure of any one component required that the entire gun had to be returned to the same craftsman who would have to create from scratch a unique and costly replacement.

In the late 1700s, a French gunsmith realized that the cost of guns could be dramatically reduced if all of the parts of a musket could be *standardized*. Standardization meant that parts from one gun would be interchangeable with the same parts with any other gun of a given design.

In 1785, a gunsmith named LeBlanc met with the US Ambassador to France, then an obscure scientist/inventor/statesman named Thomas Jefferson. He challenged the American to randomly pick parts from a series of bins and then to assemble them into a perfectly functional sub-component of a musket. Jefferson was impressed enough with this demonstration to write home about what he had seen.

The idea of standardized parts was immediately adopted by Eli Whitney, the inventor of the cotton gin. In the late 1790s he secured a contract with the US government to produce and deliver 10,000 muskets in less than twenty-four months, a task that was considered to be impossible using the existing manufacturing techniques. Rather than beginning to build muskets immediately upon getting the contract, Whitney spent several months building and tooling a musket assembly line in a new factory located in Hamden, Connecticut. When completed, the factory was equipped with a series of water powered machines, each designed to perform one specialized manufacturing task over and over again to spew out an endless series of uniform, identical and interchangeable high quality musket parts. When that plant reached full production in 1803, it was capable of producing more than 10,000 muskets per year, or three times more than could be produced by traditional skilled gunsmiths. By 1807, the improved factory reached an output of more than 20,000 muskets per year.

The concept of standardized, interchangeable parts revolutionized manufacturing. Albert Einstein once said that if he was given one week to solve a problem, he would spend five days making sure that he fully understood all aspects of the problem before devoting the last two days to formulating the solution. Similarly, in 1807 when Eli Terry accepted an order to build 4,000 clocks, he spent more than a year preparing to mass produce the devices by setting up all of the components of the assembly line. After launching, he managed to produce more than 1,000 clocks in the first year of production and over 3,000 in the second. His costs were so low that he was able to undermine the costs of every handmade clock then on the market. Between 1850 and 1900, thousands of inventions that were to make modern life possible were produced using the concept of standardized parts. These innovations included the sewing machine, escalator, elevator, air conditioner, typewriter and many more.

These inventions depended on a whole network of innovations in many different areas. For example, in metallurgy the development of new steel alloys allowed the production of lighter and stronger metal components. The development of precision machining led to higher-tolerance moving parts. While in the 18th century steam engine cylinders were manufactured to an accuracy about equal to the thickness of a coin, by the beginning of the 20th century accuracies of more than one thousandth of a centimetre could be achieved. The new assembly lines were constantly being improved by corresponding advances in the machines that were being designed to make other machines such as drills, grinders, lathes, hammers and presses. At the Philadelphia US Centennial Exposition held in 1876, almost 10,000 different machines were placed on display ranging from tiny sewing machines that weighed less than a kilogram to monster 2,000 horse power steam engines that weighed tonnes.

With spreading mass production came rapidly falling costs for manufactured goods. This immediately benefited the poor, who were able to afford what used to be "luxury" products for the first time.

Numerous innovators attempted to bring scientific discipline to the assembly line. For example, Frederick W. Taylor studied virtually every aspect of mass production including determining what was the optimum speed that tasks should be carried out, how heavy tools should be, and how far should workers have to walk and reach in between tasks. Frank and Gillian Gilbreth studied every aspect of the details of human motion, with the goal of making workers more efficient. Applying the results of scientific "Taylorism" resulted, in some cases, in a doubling of assembly line productivity.

The modern assembly line represented an attempt to integrate human beings with machines. Like the machines, the individuals manning the lines had to repeat the same actions again and again within strict time limits. With the development of increasingly capable machines such as semi-intelligent assembly robots, humans were gradually removed from the most dangerous and "dirty" of the assembly line jobs. But even though the details of the assembly lines were improved in many ways, the most modern of the 21st century assembly lines were still based on the original 19th century model integrating men and machines.

The first modern assembly line was constructed in the Highland Park Automobile Factory by Henry Ford and his production engineer, Charles Sorensen. The first vehicle off the line was the now famous "Model-T" which was an assembly that consisted of more than 5,000 standardized and interchangeable parts. When advising Sorensen, Ford told him that the line had to be designed so that no man had to take more than one step in any direction to get to the parts that he needed and that no worker ever had to bend over to reach a part.

Figure 1-3: Ford Assembly Line in 1928. Source: Literary Digest 1928-01-07 Henry Ford Interview/Photographer unknown

An example of the impact of the scientifically designed assembly line could be seen in the fact that while one skilled person required more than twenty minutes to put together a given assembly, when the job was divided into twenty-nine separate operations done by as many individuals, the time dropped to about thirteen minutes. Raising the height of the assembly line by only twenty centimetres (bringing the components into more convenient reach) reduced the time to only seven minutes. Experimenting with the speed of the line further reduced the time to less than five minutes.

Ford observed that assembling a Model-T chassis using the original stationary method required more than twelve hours. When placed on a waist high assembly line moving at two metres per minute through forty-five assembly stations, the assembly time was reduced to about two and a half hours. The work was specified right down to the last detail. Those placing the parts did not attach them, those inserting bolts did not fix the nuts and those fixing the nuts did not tighten them.

The success of the modern assembly line could be seen in the Model-T production figures. While more than 78,000 units were produced in the period 1911-1912, production rose to over 785,000 units during 1916-1917, while the price per unit fell by almost fifty percent.

Since Ford's time, assembly lines were built in every corner of the globe and regular technological advances continued to improve their efficiency and the quality of the products that they produced, but the basic principles of production remained the same.

Despite all of its great successes, by the end of the 20th century it was becoming painfully apparent that there were a number of serious problems with the traditional assembly line system. These included:

- It was designed to produce long runs of virtually identical products

- Complex and detailed specs were given to hundreds of outside suppliers who were given little room for innovation

- The failure of only one "link" in the supply chain could bring the entire assembly line to a halt

- The inability to produce custom products for individual clients

- Difficulty in responding quickly to changing consumer demands leading to the creation of large inventories of unsold products

- Prices for products were inflated by the necessity of supporting a large workforce of unionized workers with expensive benefits packages

Many attempts were made to "patch-up" the deficiencies in the existing manufacturing model, such the adoption of Just-On-Time-Delivery of parts to the assembly line, or the development of Flexible

Manufacturing (FM) in an attempt to increase the speed at which assembly lines could adapt to evolving customer tastes. These stop-gap measures could only delay, but not prevent the inevitable decline of the popularity of the assembly line model of industrial production.

Conclusion

To overcome the limitations inherent in the traditional assembly line model it was necessary to invent a completely new approach to manufacturing. This new system will profoundly change the world and will eventually touch the lives of virtually every human being on earth. One name given to this new industrial paradigm and all of its related and supporting technologies was "JOOM" or Just-On-Order-Making.

This book is not an introduction to 3D making technologies. There are many existing books and on-line resources that can provide the interested reader with all the information that they may desire on this topic. Rather, the purpose of this book is to examine in-depth how rapidly evolving 3D printing and supporting technologies will contribute to the upcoming global JOOM revolution.

The many dimensions of this paradigm-shifting revolution will now be examined.

Web Resources

Use of 3D Printing by NASA:

http://www.policymic.com/articles/62055/
nasa-tests-the-limits-of-3d-printing-with-this-incredible-project

3D Printing & U.S. Navy:

http://www.usni.org/magazines/proceedings/2013-04/print-me-cruiser

3D Printing & U.S. Navy:

http://www.popsci.com/technology/article/2013-04/future-navy-3-d-printing

3D Printing & U.S. Navy:

http://www.extremetech.com/extreme/156773-us-navy-looks-to-
3d-printing-to-turn-its-city-sized-aircraft-carriers-into-mobile-factories

3D Printing & U.S. Navy:

http://www.technewsdaily.com/18127-3d-printing-could-revolutionize-navy.html

Photo Sources

Figure 1: Dr. Paul D. Tinari, Naval War College, Newport RI, 2003

Figure 1-1:

http://upload.wikimedia.org/wikipedia/commons/thumb/b/b3/R
%C3%B6mische_S%C3%A4gem%C3%BChle.svg/2000px-R%C
3%B6mische_S%C3%A4gem%C3%BChle.svg.png

Figure 1-2:

http://commons.wikimedia.org/wiki/File:Sawing_Machine_Model.jpg

Figure 1-3:

http://commons.wikimedia.org/wiki/File:Ford_
Motor_Company_assembly_line.jpg

Bibliography

Ashton, T.S. (1948) "*The Industrial Revolution, 1760-1830*," Oxford University Press

Brady, R.A. (1961) "*Automation & Society: The Scientific Revolution in Industry*," University of California Press

Burlingame, R. (1953) "*Machines that Built America*," Harcourt, Brace and World

Derry, T.K., Trevor, I.W. (1961) *"A Short History of Technology,"* Oxford University Press

Dunlop, J.T. (1962) *"Automation and Technological Change,"* Prentice Hall

Henderson, W.O. (1961) *"The Industrial Revolution in Europe, 1815-1914,"* Quadrangle Books

Lilley, S. (1948) *"Men, Machines & History,"* Cobbett Press, London

Mantoux, P.J. (1961) *"The Industrial Revolution in the 18th Century,"* Macmillan

Usher, A. P. (1959) *"A History of Mechanical Inventions,"* Beacon Press

White, L. Jr. (1962) *"Medieval Technology and Social Change,"* Oxford University Press

Wiener, N. (1950) *"The Human Use of Human Beings,"* Houghton Mifflin

Chapter 2
The First JOOM Revolution
Just-On-Order-Manufacturing

Introduction

Leonardo da Vinci was probably one of history's greatest creative geniuses. He invented dozens of engineering contraptions including different types of flying machines, the armoured tank, submarine, various industrial devices, and the assembly line. But as far as was known, all of his inventions remained on paper. The opportunity to build prototypes and test his theoretical concepts with real-world 3D models was never realized.

It is now possible for anyone with access to a rapid prototyping/3D printing machine to make a real-world model of virtually any new concept with little more than the touch of a button. As an example, a fluid dynamics expert can design a completely new type of propeller using CAD software and then can insert the 3D model into a computer-based, virtual water tunnel to test its performance characteristics under different flow conditions. When deficiencies were identified in the design, it could be popped back into the CAD program to have the design modified, retested, reconfigured and then tested again to confirm that it now operated as expected. When the scientist was completely satisfied with the performance of the component, the final design could be sent to a 3D printer to be produced and the designer could be confident that it would perform exactly as expected.

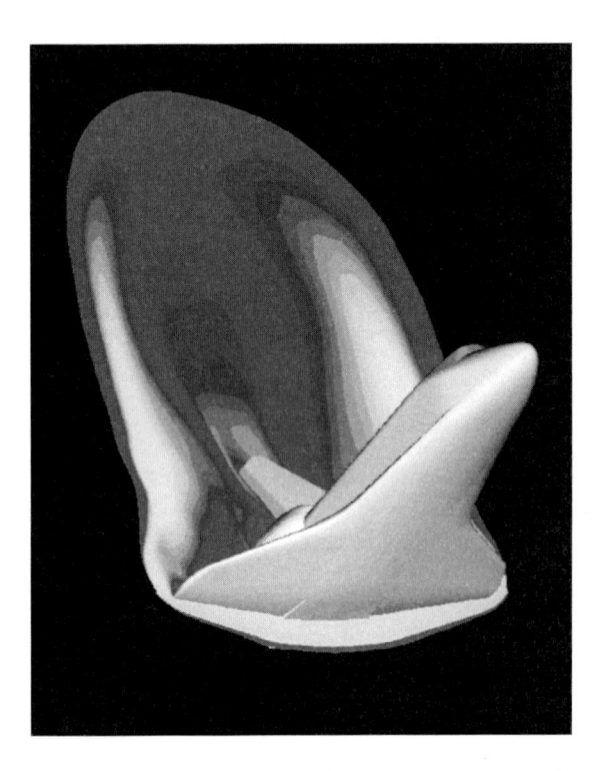

Figure 2-1: CFD Simulation of the flow over an orbiter prototype in a virtual wind tunnel. Source: NASA

A number of books have examined 3D printing and various supporting technologies. But few authors have considered in any depth the profound impacts that these developments will have on all aspects of modern society. Before an examination can be made of these impacts however, it is necessary to first gain some sense of the history behind the critical technologies associated with 3D making.

The Assembly Line

Perhaps none of Leonardo's inventions had a greater impact on the world than the assembly line. The modern assembly line is now over a hundred years old. In the years since it was first perfected by Henry Ford, there have been many improvements to the standard assembly line such as the introduction of Just-On-Time-Delivery (JOTD) of components, the use of Total Quality Management (TQM) and the gradual introduction of robots and other forms of automation to replace human workers in the really gross jobs. But all of these represented just a "tinkering" with the operation of the system and not a fundamental re-thinking of the whole industrial model.

The massive Ford Rouge assembly plant represented the highest achievement of the philosophy of vertical integration. At one time it made perfect sense: before the development of networked communications, it proved impossible to coordinate deliveries from many independent small suppliers. Logically, it made sense for Henry Ford to set up a system where he had total control. That control extended to raising the sheep needed to produce the wool for the seats in the cars that were to be produced on his assembly lines. However, after sixty years of operation, the limitations and dysfunctionalities of this rigid, inflexible system, including its lack of creativity and its intrinsic slowness to adapt to new conditions became increasingly apparent. However, what seemed obvious to almost everyone outside of the automobile industry remained mostly a mystery to its top executives. This puzzling inaction in the face of overwhelming evidence was explained by Prof. Clayton Christensen of Harvard University in his book "The Innovator's Dilemma":

> *"Successful companies in mature industries rarely embrace disruptive innovation because, by definition, it threatens their established business models. Loath to revamp factories at high cost to make products that will compete with their own goods, companies drag their feet;*

perversely, financial markets often reward them for their short-sightedness."

Corporations using the traditional assembly line system continued to face, just as they have always faced, the following problems: How many of each item should be produced and what specific characteristics should those items have? How could the retooling of each assembly line be sped up so that new features could be incorporated into products in a minimum amount of time? How could customers be assured that they would get exactly the specific characteristics that they really wanted in each product and none of those that they didn't want? How was it possible for an organization to remain competitive when it had to deal with a highly militant workforce mostly made up of relatively unskilled workers who were obstinately resistant to any real change? Each of these critical questions will be examined here in turn.

It was very difficult to predict the quantities of each item to produce. For example, how many red cars should an automobile company produce? How many blue cars? How many four cylinder cars vs. six cylinder ones? In fact the possible questions were literally endless. Traditionally, organizations arrogantly believed they knew what customers wanted and they simply designed, produced and delivered it. This explained Henry Ford's famous comment, "They can have a car in any color they want as long as it is black."

Many years later, the Ford Motor Company undertook what was then considered to be a revolutionary approach to developing a new automobile by spending many hours actually asking real customers what they really wanted to have in the car that they would buy. This mass of data collected from prospective car buyers was analyzed and used to guide the design engineers. The result was the Ford Taurus, which went on to become one of the best-selling American automobiles of all time. This approach was gradually adopted by other corporations, as it now seemed to be a common-sense notion that if you really wanted to know what customers wanted, then you simply asked them.

Figure 2-2: A view of a typical modern assembly line. Source: Siyuwj

The shutting down of an assembly line for retooling every time that it was desired to produce a new model of a particular product was a mandatory process for virtually all modern assembly lines. Depending on the complexity of the required retooling, the length of the downtime could range from a few days to several weeks. Japanese companies gained a significant competitive advantage by dramatically reducing the time required to retool their assembly lines.

Consequently, organizations all over the world subsequently worked diligently to minimize the duration of the periods of zero production, but it proved impossible, while working within the limitations imposed by the assembly line paradigm, to eliminate the need for retooling altogether. For over a hundred years, companies increased their competitiveness by addressing the various inefficiencies intrinsic to the traditional assembly line. However, technological developments reached the stage where a radically new approach could be taken to the production of manufacturing goods, one that would immediately make the assembly line and all of the systems associated with it as obsolete as buggy whips.

In 2011, the Ford Motor Company approached the government of Canada for financial aid to reconfigure its auto assembly plant in Oakville, Ontario. This request was made after the government awarded Toyota more than $140 million to upgrade its plants in nearby Cambridge and Woodstock. This help was in addition to the $100 million that the government gave the Ford Motor Company in 2006 to "rebuild" its Oakville operations. And this was on top of hundreds of millions of additional taxpayer money given to the auto industry over many decades. In the US, the state of Tennessee offered Volkswagen more than $500 million to locate a new assembly plant in that jurisdiction. During the 2008 recession, the government offered more than $13 billion U.S. to Chrysler and General Motors to help them avoid default. A cynical view of all of this government generosity to one specific industry was that the members of the United Auto Workers union had become little more than government employees. Perhaps a more accurate observation was that long established companies such as the Ford Motor Company were attempting to co-opt politicians to help sustain a dysfunctional business model that could only be kept operating by the endless injection of countless millions of taxpayer-funded dollars.

To their credit, the large manufacturers introduced many innovations to try and sustain the assembly line model. Examples included: Six Sigma, Total Quality Management, Just-On-Time (*NOT* the same as Just-On-Order), Sequenced Parts Delivery and Synchronized Manufacturing. (Eisler et al. 2007) However in the end, the sad realization that many auto executives had to accept was that all these strategies had become obsolete with the introduction of what would become the next great paradigm of manufacturing: the era of Additive Manufacturing.

To succeed in the 21st century, organizations had to forget everything they thought they knew about product life cycles, marketing, product design, engineering and manufacturing, and had to be prepared to start from scratch with a new technological paradigm. Traditional approaches to prospecting for sales were highly inefficient. Most old-style vendors had no idea how their customers were using new technologies, so their efforts to connect with them often failed. Technology transformed the one-way messages of the mass marketing efforts of the 20th century into the real-time marketing dialogs of the 21st century. This new market consisted of a virtually infinite number of niches, some of which were only occupied by a single individual.

Consequently, web marketing was all about delivering personalized content at the exact time that the purchaser wanted it. With the increasingly global reach of social media, people could now get the information they needed about products and services that interested them through their increasingly intelligent "smart" phones, not only from vendors, but of greater interest to them, from people they knew and trusted. The role of vendors evolved from simply supplying undistinguished products and services that simply solved the buyer's problem(s) to entertaining, exciting, delighting and enchanting them. Successful organizations were the ones that had made the shift from "telling & selling" to actively building relationships with customers. All successful business models had to be first and foremost customer-centric ones that maintained relevancy by connecting contacts with quality customized content.

Steve Jobs once said that the old assembly line industrial paradigm was so slow that if your business model was based on asking customers what they wanted, by the time you got it designed, manufactured and into their hands, they would have already moved on to wanting something else.

Twitter and other social media were not simply seen as technologies or channels to be used by vendors to push out information. Rather, they were seen as a conversation with an invisible audience. This conversation was going to happen whether a particular organization wanted it or not. Buyers could reach out to a vendor's site and search for far more than a particular product or service, but would also search for ways to better understand themselves, their world, and their place in it.

Historically, many organizations invested millions of dollars each year into the development of new rules. Unfortunately, all they were succeeding in doing was spending to perpetuate mediocrity. To be successful, 21st century organizations had to continuously learn from other's failures so they could then invest in new *ideas*. Whenever an old paradigm business model was made obsolete by a revolutionary new technological idea, those formally dependent for their survival on the old model would do their utmost to prevent the deployment of the new technology. Don Tapscott, the author of "Wikinomics", said: "Holding back technology to save a broken business model was like letting blacksmiths ban the internal combustion engine to protect horseshoes." Large centralized organizations such as the American "Big Three" auto makers prospered for generations with their tried and true business model (along with a great deal of financial and political help from governments), were now living on borrowed time and the sad thing was that they were not even dimly aware of this fact.

The opening years of the 21st century saw a string of industrial closures caused by intense global competition. In central Canada alone, Caterpillar, AstraZeneca, Navistar, the Ford Motor Company, 3M and many other companies either terminated or significantly reduced the scale their industrial operations. (Globe & Mail 2012) Traditionally, heavy industries would shift operations to regions with significantly lower wage rates such as China, Bangladesh or Vietnam, but with increasing automation, wage rates played an increasingly insignificant role in corporate decision's to restructure, shut down or relocate. Rather, by the second decade of the 21st century, the future of industries such as the automobile sector was becoming clear to a number of analysts. Typical were the words of Charles C. Mann, who wrote; "Ultimately, modular construction will lead to cars that are custom built to the specifications of their future owners. " (Mann 2009)

Late in 2011, Research in Motion (RIM) announced that its "finished goods" inventory, referring to products such as its flagship BlackBerry and Playbook devices, sat at almost $300 million, acting as a massive drag on the entire company. The flaw in their business model was that it booked revenue for its products when they were sold to retailers and wireless carriers and NOT when they were actually sold to the paying end users. Clearly, a competitor attempting to move into RIM's marketplace could do so by abandoning the "build it and store it in inventory if it does not sell" business model and moving to the far more powerful model of "build it just when you have a paid order" model. The proven effectiveness of this new model, enabled by the development of 3D printing and related technologies, will eventually sweep all other less efficient models into the dust bin of history.

Also in 2011, the giant French company PSA Peugeot Citroen experienced a severe shortage of screws caused by operational problems at the supplier factories. Without screws, the company was unable to mount shock absorbers, bumpers and other critical automobile components on its assembly lines. In the same year, massive floods in Thailand combined with the earthquake and subsequent tsunami in Japan, painfully demonstrated the fragility of global supply chains. The subsequent analysis of these events indicated that the push for lowest-cost parts made in foreign factories, lean supply chains and just-on-time parts delivery and manufacturing had significantly increased the vulnerability of major manufacturing concerns to supply disruptions. Clearly what was needed was an *on-demand* manufacturing approach, to make all of the screws and parts required for the items on the assembly lines only as they were needed, and not a moment before.

And what was a possible name that could be given to this revolution in technology that would eventually prove as critical to human society as the discovery of fire? One possible name that effectively encompassed the essence of this revolution was JOOM or Just-on-Order-Making.

An examination will now be made of the various enabling technologies that made JOOM possible.

Early JOOM

By the second decade of the 21st century, professional journals were full of articles reporting on the use of CNC (Computer Numerical Control) machines to manufacture small parts to high precision from materials that were traditionally hard to cut such as stainless steel, nitinol or titanium. High-resolution laser processes were also developed that were able to accurately create features less than a micron (0.001 mm) in size. The great advantage of this approach was the ability to work with a broad spectrum of materials including most pure metals, metal alloys, ceramics and plastics such as PET, PEEK, PEBA, PMMA, ABS and many others. (MPMN 2012)

Figure 2-3: A typical CNC machine for hobbyists and the DIY market. Source: ShakataGaNai

For a very long time the U.S. Department of Defence did not notice that there was an exploitable loophole in its regulations. All bills for shipping spare parts to active combat areas or to U.S. military bases that were labelled "priority" were paid automatically, without oversight. A typical example of how private firms could exploit this was the case of C&D Distributors in Lexington, S.C., which sent a bill to the government for $998,798 for shipping two nineteen-cent washers. In another incident, the Pentagon paid $455,009 to ship three screws costing $1.31 each to Marines in Iraq. (Globe & Mail 2007) This situation so upset the highest levels of the U.S. military command that the decision was taken to make it an urgent priority to seek a solution that would guarantee that the military could never again be held hostage by its need for spare parts.

A few years later, it was reported that the earthquake, tsunami, and nuclear crisis in Japan caused severe shortages that seriously affected North American automobile production. At one point it was estimated that critical shortages of Japanese-manufactured spare parts and components could cause an incredible forty percent reduction in global automobile production. (Globe & Mail 2011) Even though the Pentagon did not admit it, the parts shortage also affected the military, but in the intervening years since the C&D incident, steps had been taken to assure a ready supply of spare parts to all military units. One word could be used to summarize the "steps" that had been taken by the U.S. military to assure an uninterruptible supply of spare parts: JOOM.

A number of entrepreneurs became frustrated with the high costs and long lead-times traditionally associated with the production of custom designed injection moulded parts. This situation offered a perfect business opportunity for organizations such as Protomold and the Moldflow Corporation that promised to deliver injection moulded parts on short time scales and at low costs. Once an order was submitted in the form of a 3D CAD design created in one of a number of recognized formats, the approach was to use software that automated the mould design and CNC tool-path generation process, allowing parts to be rapidly manufactured and shipped. It was true that early 3D printers produced products with inferior material properties and with very rough service finishes, so injection moulding certainly offered a distinct manufacturing advantage. But the technology continued to evolve, and new generations of 3D printers largely overcame the limitations that once made them a secondary choice for entrepreneurs wishing to make parts for their experimental prototypes.

In 1984, Charles Hull developed one of the most successful additive manufacturing technologies when he realized that ultraviolet light could be used to selectively solidify a light-sensitive plastic. The technique that he developed was to have a movable platform positioned in a vat of liquid resin. The UV light was used to solidify a microscopically thin layer of the liquid over the platform. After each layer was completed, the platform would move down by the width of the solid layer and fresh liquid would flow over the solidified resin, ready for another layer to be solidified into the computer-defined form for that specific layer. Mr. Hull later went on to found 3D Systems, a company that would go on to be a leader in the rapidly evolving field of 3D Stereo-Lithography. (Lemley 2000)

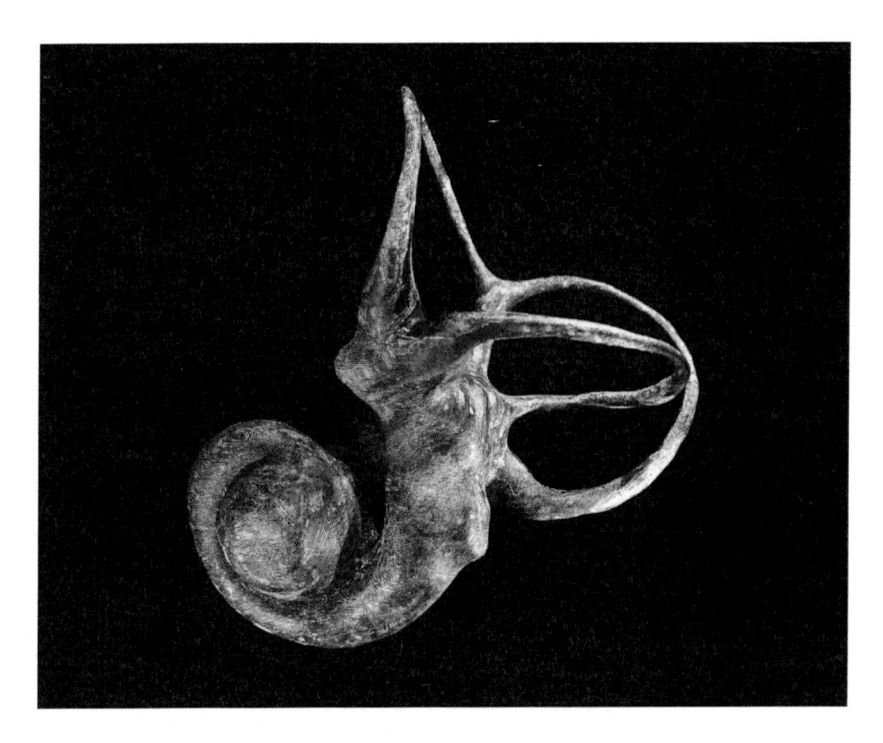

Figure 2-4: What a Stereolithograph machine can do. Model of the Inner Ear of an Extinct Primate. Source: Didier Descouens

In the 1990s, CAD software reached a level of sophistication that allowed Virtual Prototyping (VP) to enable designers to create every component of a complex product such as an automobile, and then to conduct detailed engineering analysis of the performance of each component of the final assembled product, without ever having to build it. A virtual car could be so accurately represented that it had an almost holographic quality. The virtual car could be driven on a virtual test-track, have its drag coefficient evaluated in a virtual wind tunnel and have the effectiveness of its virtual headlights tested on a virtual foggy night. Huge time savings could be achieved in new product development since traditional Finite Element Analysis (FEA), Computational Fluid Dynamics (CFD) and Thermal Analysis (TA) could be conducted on the virtual prototype long before the final assembled product was actually built.

Researchers at the Massachusetts Institute of Technology (MIT) developed a technology where a 3D printer deposited a binder solution through a nozzle onto a plaster-based powder. In post-processing, the material could be infiltrated so that it would have properties similar to rubber, plastic, or even chrome. The technology could also be used to create casts and moulds that were used to make metal or urethane parts. In 1994, MIT created the Z Corporation to commercialize this 3D printing technology. In 2001, the company was recognized as being the first to produce a 3D printer capable of producing coloured components. (Z Corp 2005)

In the opening years of the 21st century, physicist and professor Neil Gershenfeld, director of MIT's Center for Bits & Atoms, had the idea of creating a compact factory that could be fitted into a standard shipping container and that could be easily and quickly delivered anywhere in the world. His concept of a mobile Fabrication Laboratory (Fab-Lab) was a platform that included a laser cutter, CNC wood router, miniature mill, lathe, industrial press and more, that could be used by relatively untrained people to manufacture virtually anything, all for a cost of less than $50,000. Users of Fab-Lab were encouraged to "dream-it, make-it and then to share-it." True to their original concept, Fab-Labs were shipped all over the world to help people fulfill their visions. In Afghanistan, a Fab-Lab was dedicated to making prosthetic limbs for child victims of anti-personnel mines. In South Africa, the system was harnessed to make simple, durable, and inexpensive computers; in Norway, it was applied to create RFID (radio frequency ID) tags to keep track of sheep, while in the South Bronx, one was used by ghetto children to make modular toys out of waste cardboard. In the words of the inventor and developer of the Fab-Lab concept: "If anyone can make anything anywhere, it fundamentally changes the meaning of business." (Greenberg 2008, Okafor 2011)

Figure 2-5: A 3D Printer typically used in a FAB-Lab
installation. Source: Benoît Prieur

It was well known to the major clothing manufacturers that customers were becoming increasingly frustrated with the huge discrepancies in clothing and shoe sizing. A company named Unique Solutions Design Ltd. developed the Me-Ality Scanner, a machine that used low-energy radio waves to generate 200,000 data points mapping out detailed measurements of the body of a client. The software would then generate the sizes that the person should choose for each brand and style of clothing for the best possible fit. A German-based company called UPcload allowed users to dispense with the scanning booth by developing a web-based scanning system that used a client's computer web-cam to generate measurements that were almost as accurate as would have been taken by a professional tailor. Intel developed an "app" known as Magic Mirror that, when supplied with measurements provided by a client, created an avatar that could be used to try on a wide variety of clothes to evaluate them for form and fit. The software could be used to match complete sets of accessories to the clothing to make virtual outfits, all shown with helpful suggestions about where they could be purchased. (Cooper 2012)

In the 1990s, a non-profit research corporation known as the Textile/ Clothing Technology Corporation (or TC^2) began researching the possibilities of creating custom tailored clothing for every customer. The procedure that they eventually developed required the client to strip down to a pair of tight-fitting briefs. The person was then scanned by laser triangulation until the positions of more than a quarter million points on the surface of the body could be established in 3D space, creating a "cloud" of data points. Advanced software joined together these data points to form a 3D "mesh," or surface, smoothed and simplified it, and then finally extracted all of the measurements relevant to the design of clothing. (Siscia 2008)

The data thereby created was so accurate that it could be used to design virtually any item of clothing, from a form-fitting waistcoat to a wetsuit. After all the body measurements were completed, the client selected from a tablet menu all of the personal details of the clothing including if they wanted pockets or custom monograms woven into the shirts. After the order was completed, all of the information was sent to the computers controlling the manufacturing machines and in a very short time the client was handed a neatly folded pile of perfectly-fitted clothing that was ready to wear in what could be termed a pure manifestation of Just-On-Order-Measurement-and-Manufacturing. (Lajoie 1999)

A Montreal-based firm with the rather futurist name of M0851 decided that, in order to remain in North America, it had to dramatically change its business paradigm. It consequently invested more than half a million dollars into an autonomous robotic leather cutting machine that could dramatically increase manufacturing throughput without reducing quality. Interfacing the machine to the internet allowed technicians in Italy to make continuous adjustments to the machine while it operated in Canada. The great advantage of using the machine was the ability to almost instantly adapt to rapidly changing fashion trends. (MacGregor 2012)

HyperStealth was a company that developed and manufactured custom-designed camouflage patterns for specific military and other applications. The company obtained a digitized printing system that could take patterns composed on a computer graphic design program and print them directly onto fabric. Short-turn orders could be manufactured just as they were ordered, meaning the company could usually beat all their competitor's prices because it carried absolutely no inventory.

A company called Military Wraps developed a proprietary site-specific camouflage technique that combined photographic digital detailing with vinyl-adhesive wraps that were designed to match a surrounding terrain in each area so exactly that anything covered with the camouflage simply disappeared into the surrounding environment. (Military Wraps 2014)

The next generations of camouflage clothing did not have one pattern printed onto their surfaces, but had an electronic-powered fabric connected to a power supply that permitted the user to change the pattern at will. Slightly more advanced was a uniform that sensed its environment and automatically changed the camouflage pattern to match, just like a chameleon, to produce Just-On-Order-Invisibility. Beyond that, the company was experimenting with a technology known as quantum-stealth that used a proprietary light refracting material to bend light completely around objects, making them, if not invisible, at least very difficult to see. (Clarke 2011)

In 2005 the Israeli firm Object Geometries Ltd. developed a process known as photo-polymer jetting. The raw material was shipped in safe, sealed cartridges and as each layer was deposited by the 3D printer nozzle, it was cured by exposure to a UV light source. Consequently, the user was never exposed to the liquid resin. The process offered a resolution of 16 microns (just a little larger than two red blood cells), resulting in a high quality surface finish. Models could be constructed as quickly as the head of the 3D printer could move over the surface. (Advanced Manufacturing 2005)

An early attempt at JOOM was known as "co-creation." The premise of co-creation was that since the pace of innovation was so rapid, to be successful in the 21st century, organizations found that they had to move away from inventing and developing new products by themselves, to embracing a new business model that involved clients in each stage of the product development cycle. It had become obvious that to provide a custom design and a unique experience for as many customers as possible, the clients themselves had to be solicited to provide the bulk of the input required to define all of the details of the design of a given product. Increasingly, when an organization such as Apple constructed a platform such as the iPad, it was the users that provided virtually all of the content.

In the case of life insurance, it was possible to create a system where the yearly cost of medical or life insurance would continuously vary according to a number of personal parameters that could be monitored such as age, diet, medication, degree of exercise, lifestyle and risk factors such as smoking, drinking and recreational drug use. The problem was that most existing organizations were simply not organized in a fashion to capitalize on the benefits of co-creation. It became readily apparent that if they were to survive, they had to reinvent themselves or rapidly lose market share to companies that could commit themselves to co-creation. (Colvin 2008, Prahalad et al. 2008)

In the opening years of the 21st century, a number of organizations introduced what they termed Quick Response Manufacturing (QRM) as a corporate strategy that sought to reduce lead times in all of the company's operations. As seen by customers, QRM meant that their requirements would be responded to by quickly creating and then manufacturing items customized to their specific needs. Some organizations implementing QRM were able to achieve reductions in turn-around times of up to ninety-five percent, cost reductions of between fifteen and thirty percent, improvements in product delivery performance of up to ninety-nine percent, and a dramatic reduction in waste and requirements for re-work by eighty percent. (Suri 2001)

A number of new start-up sites attempted to implement various components of Just-On-Order-Making. ShoeDazzle custom designed shoes for clients based on a questionnaire that they filled out, however they actually manufactured the shoes in main-line factories. Other sites included Honest for "organic" diapers, Birchbox for cosmetics, Wittlebee for children's clothing, JustFab for shoes and bags, BeachMint for just about everything, and many, many more. The weakness of the business model of all of these sites lay in two general areas. The first was that they depended on convincing customers to pay monthly subscriptions to generate long-term income. This assumption proved to be rather weak. Just as information had the universal need to be free, even for custom made items consumers wanted to be free to shop when and where they wished. The second area of weakness was that each of these businesses was highly capital intensive because their model had them not only designing, but also stocking and shipping their own inventory, which was essentially the Internet equivalent to putting propellers on a Boeing 787 Dreamliner and expecting it to fly better. They had all violated the essential philosophy of JOOM, which was to strive to have no inventory. (Miller 2012)

It took a large mainline company like the Ford Motor Company from two to three years to plan, develop and build a new automobile. When creating a new bumper design, because of the preparation involved in reassigning the stamping machinery and in properly integrating it into the assembly line, the manufacturer had to "lock-in" the design months before the first automobile using the new bumper design could roll off the assembly line. Traditionally, the design of a vehicle was considered "locked in" from the moment it was produced by the manufacturing plant to the moment it was deposited into the wrecking yard. The whole basis of automobile recalls was that any mistake during the design was extremely expensive to correct after the product was delivered into the hands of the customer.

In 2011 the Ford Motor Company mailed out memory sticks to more than a quarter million vehicle owners so they could update the software in the vehicle's touch-screen. This was considered to be a relatively primitive form of Just-On-Order-Software. The next step was to allow the vehicle computer to wirelessly upgrade automatically whenever necessary, so that the systems could be placed into a state of perpetual renewal, just as Internet-connected tablets were being wirelessly supplied with the most recent software upgrades.

In an attempt to impose some flexibility on the traditional assembly line, General Motors redesigned and then reopened its shuttered Saturn manufacturing facility in Spring Hill, TN. The facility was gutted and then rebuilt from the ground up to be a so-called "flexible" plant. The meaning of flexible in this application meant the ability to efficiently and economically assemble relatively small numbers of vehicles for short periods of time. The problem that this new plant was meant to solve was to be able to ramp up quickly to immediately produce popular vehicle models without having to invest in new tooling, assembly lines, or plants. A flexible assembly line was able to rapidly switch to another vehicle model as soon as the market popularity of a successful car began to fade. This change made sense because of the need to adapt to the rapidly fragmenting auto market of the 21st century, where often less than 30,000 units of a new model type could completely saturate a market segment.

In the same vein, the Ford Motor Company attempted to redesign its Focus model to be both modular and global. It was modular in the sense that the electric, diesel and gasoline models of the Focus could all be assembled on the same assembly line. At the same time, it was global in the sense that a change to the vehicle design in one part of the world could be instantly incorporated into the models being assembled in all of the other plants around the world.

The recession of 2008 hit a lot of manufacturers hard and many were able to survive *only* by breaking the old business models that had once served them so well. One new idea first tried by the pharmaceutical industry was modular manufacturing, where the process was divided into discrete, standardized units that could be easily assembled/disassembled so that they could be quickly moved to any desired location in the world. All that was required to start operations was the connection of the assembled modules to utilities such as power and water. Larger, more complex processes could be constructed of increasing numbers of interconnected modules. Systems could be quickly designed to produce a facility of any required size with reduced lead times, higher productivity and lower set-up costs. Since each module was completely tested, this approach also offered greater reliability, increased flexibility and reduced production costs. A change of process required only the disassembly of the system and then a rearrangement of the modules in the configuration required for the new process. (Greb 2010)

A company in the unassuming town of Wareham, Massachusetts, was one of the first open-source car manufacturing operations to reach full production. In June of 2010, Local Motors as it was called, released an off-road (but street-legal) racing car called "The Rally Fighter," for about $50,000. What made this particular car design unique was that the design was Crowd Sourced. When the customer purchased a vehicle, it was not fully-assembled, but was delivered as a number of boxes of components along with some assembly instructions. The company aimed to take a new vehicle from CAD concept to a market-ready kit in less than eighteen months, or about the about the time it would take a traditional automobile manufacturer to design a new cup-holder. Another unique feature was that each of their designs was not considered to be proprietary, but was released under a user-friendly Creative Commons License. Customers were actually encouraged to enhance the company's designs and to produce their own improved components that they could in turn sell as they wished.

Figure 2-6: The Rally Fighter vehicle made by Local Motors.
Source: http://www.flickr.com/photos/markus941/

In the past, kit vehicles were typically modeled after famous racing or sports cars, often inflating the cost of the kits with annoying lawsuits and license fees. Local Motors overcame this problem by only selling original designs. With increasing numbers of gifted amateur designers being well-equipped with leading edge 3D CAD software packages combined with photorealistic rendering technology, the result was the creation of some truly outstanding original vehicle designs.

The company kept the creative juices of its contributing audience of designers flowing by periodically running design competitions with significant cash prizes. While the community concerned itself mostly with applying its significant group creativity to the design of the exterior of the vehicle, the company professionals also worked on the components critical to safety, structural stability and manufacturability of the car such as the engine, transmission and structural framing.

Local Motors never needed more than a dozen full-time employees to manage an output of more than 2,000 vehicles of each type and it accomplished this while sitting on almost zero inventory. Also, it produced all of the components for a particular kit only after buyers paid for it, or at least had handed over a substantial down payment. The company officials realized that they were operating in a relatively small niche market and that they would never be seen as a threat by the "Big Three" auto makers. But as the idea of custom-designed, highly creative, just-on-order vehicles spread, it was expected that sales volumes would continue to grow.

By 2011, many product manufacturers were hurt by the rapid escalation of domestic and international shipping costs. Cash-strapped governments endlessly piled on additional port handling fees, customs tariffs, carbon taxes, green charges and rising fuel surcharges resulting mostly from increasing government energy taxes. The time was ripe for launching new business models that bypassed product shipping and other artificially-imposed charges altogether. It also had to be capable of delivering products directly to customers.

The Rise of Rapid Prototyping

Rapid Prototyping (RP) was known by many terms depending on the actual technologies involved including Solid Freeform Fabrication (SFF), Digital Fabrication (DF), Automated Freeform Fabrication (AFF), 3D Printing (3DP), Solid Imaging (SI), Layer-Based Manufacturing (LBM), Laser Prototyping (LP), Additive Manufacturing (AM) or just simply the term favoured in this book, Just-On-Order-Manufacturing, or more generally, Just-On-Order-Making (JOOM). The term "3D printer" was used to refer to the actual machines creating the objects, while "3D making" was used as a more general term, referring to 3D printers and all of the supporting technologies such as 3D modelling, CNC machines, 3D scanners and so on.

The author first used the term "JOOM" in a ground-breaking article that he wrote for a small, Vancouver, British Columbia based business magazine. This article was one of a series that examined both the positive and negative impacts of changing technologies. It was stated that the essence of JOOM was that all manufactured products could be custom designed and were only built when a paid order was placed by the customer. (Tinari 2000)

In this work, JOOM was used to encompass all the descriptive acronyms used in the industry as well as all of the related and supporting technologies. Additive Manufacturing/3D Printing could be considered to be a form of Virtual Manufacturing, where all of the really heavy creativity and design work was done in the virtual world of the computer before anything was ever built in the "real" world. (Del Ciancio 2006)

By the second decade of the 21st century, a number of astute observers could already see that 3D making was going to have a profound impact on the world. Using increasingly affordable 3D printers, entrepreneurs could create virtual models of their proposed prototypes using 3D computer aided design (CAD) software, refine the designs by running them through a Finite Element (FE), Computational Fluid Dynamics (CFD) or other engineering package, submit the design to be patented and then create working versions of the product for consumer testing, all without leaving their basement home office. In the past, inventors had to fork out thousands of dollars for tooling, moulds, dies, machining and craftsmanship so that the first prototypes could be manufactured. As the costs for all types of 3D printers continued to plummet, they could increasingly be used by ordinary, non-technical people to create and manufacture replacement spare parts, toys or tools whenever and wherever they needed them. (Pilieci 2010)

Rapid prototyping printers not only made product availability more convenient for companies but also for consumers. The technologies once used only by large manufacturers were now available for small business or home use. Rapid prototyping printers provided an enhanced ability for the production of models and products. It represented the fulfillment of the promise offered by Demand-Driven Manufacturing. (Jakovljevic 2007) If a new brush or comb was desired, rapid prototyping printers could produce one very conveniently. Rapid prototyping printers could be used for a variety of products constructed of a broad spectrum of different materials. The inputs used by rapid prototyping printers were 3D models made by CAD software or produced from a real object that was digitized by a 3D scanner. The models were then manufactured layer by layer until an exact reproduction was generated according to the specifications dictated by the program.

If a small business needed a product or model produced, there was no longer any need to maintain workshops with specialized tools manned by expensive skilled craftspeople as would have been necessary with traditional modeling. Rapid prototyping printers will provide the technologies necessary to produce 3D models more conveniently, in shorter periods of time and for less money. In today's dynamic business environment, for a small business to prosper it had to have the capability of rapidly producing innovative and competitive products in a cost efficient manner. Rapid prototyping provided the technology for small businesses that were seeking a more convenient and cost effective manufacturing option.

One example of a rapid prototyping technology was known as Two-Photon Fabrication. The technique was to illuminate a volume of transparent, photosensitive, liquid material (usually a polymer) with an ultra-fast (femtosecond or a billionth of one millionth of a second) laser. Those volumes of material where the laser was focused became polymerized, while adjacent volumes remained liquid. By moving the position of the laser focus volume through the material, it was possible to create virtually any desired 3D shape. On completion of the shape, the non-illuminated material was removed to reveal the completed 3D figure. The great advantages with this technique included the possibility of producing objects with extremely high resolution, (less than 100 nanometres or billionths of a metre in many cases), the lack of shrinkage or distortion of the material during construction, and the fact that the technique could be used for both positive and negative photoresists. In negative photoresists, exposure to the laser activated a polymerization reaction that made the exposed area *more resistant* to the solvent that was used to remove the unexposed material. With positive photoresists, exposure to the laser led to a breakdown of the polymer chains, causing the exposed volume to be *more soluble* to the washing solvent than the unexposed areas. (Nanoscribe 2014)

The disadvantages with the technique included the high costs associated with the required high speed lasers, the relatively small sizes of the fabrication volumes, the requirement for very accurate piezoelectric positioning systems and the relatively low reliability of the entire system. However, technological advances were reducing the costs associated with high speed lasers, making this approach to 3D manufacturing of complicated shapes increasingly attractive for making, for example, implantable Micro-Electronic Mechanical Systems (MEMS) and scaffolds for tissue engineering. Studies showed that the presence of micro-meter structures in scaffolds, which could easily be produced by high resolution 3D printers, could strongly influence cell survival, reproduction and functioning. (Farsari et al. 2009)

By the second decade of the 21st century, the number of cost-effective applications for rapid prototyping was multiplying like mushrooms after a spring rain. The potential applications of 3D rapid prototyping technologies were far more extensive than traditional prototyping methods. Traditional methods required large, bulky and sophisticated equipment that also demanded a major financial investment on the part of the businesses wanting to use them. In contrast, the newest rapid prototyping printers were reasonably-sized and much less expensive. The short set-up times and simple operation of modern 3D printers made them popular for creating models, replacement parts and toys.

Inventors could now sit down with a tablet PC and sketch design ideas with the help of a CAD program. These programs allowed changes to be made to any complex 3D design as easily as words could be changed in a word processing program. Once the inventor was satisfied with the design, it could then be sent to a rapid prototyping machine that would create an exact 3D rendering of the object. As the technology of rapid prototyping advanced, the complexity and the maximum size of objects that could be made increased rapidly. At the same time, the variety of different materials that could be used for the print was undergoing a spectacular expansion. As examples of the versatility of rapid prototyping, it was possible to print out, equally well, completely functional Velcro™ strips or strips of living human tissue assembled from properly-programmed stem cells.

Rapid prototyping was a revolutionary and powerful paradigm-shifting new technology with a wide range of still largely unexplored possible applications. The process of prototyping involved quickly building up a prototype or working model for the purpose of testing various design features, ideas, concepts, functionality, output and performance. The user was able to give immediate feedback regarding the prototype and its performance. Rapid prototyping was an essential part of the process of system design and was considered to be critical to reducing project costs and development risk.

Following trends in information technologies, audio and video, manufacturing was being digitized on many different levels. Most effectively, manufacturing was gradually being combined with social media to apply social intelligence to the creation, design and manufacture of innovative new products. Within the next few years, this capability will be introduced into virtually every home. It was now possible to move beyond the era of faceless, assembly line-based, mass manufacturing to the new paradigm of personalized, home-based, custom manufacturing.

A Baltimore-based company named UnderArmour experienced great success because it harnessed the significant power of new technologies to create innovative new products. When the company engineers were designing a new shoe, they first used 3D design software to construct a virtual prototype of the item so that every detail could be scrutinized before the first real-life prototype was constructed. Then a 3D printer was used to produce a polymer version of the sole and other components of the shoe, so a detailed analysis could be made of the proposed design. The exact biometric data from the future owner of the shoe was used to custom design the supporting structure so that it would provide optimum support and shock absorbency. Together, these innovations saved more than six months and tens of thousands of dollars in development costs, giving them a significant advantage over the competition. (Fortune 2009)

The New York based company, Quirky, was one of the first to set-up an on-line collaborative community that was designed to bring innovative products through the entire development cycle, from original creative idea, to final commercialization. Quirky was established to solve one particular problem: Since the beginning of history, there have always been scores of creative individuals who were capable of generating potentially paradigm-changing ideas, but who were also frustrated by the fact that they had no means to bring these ideas to commercial success. Proponents said in support, that the site provided "engagement, interaction and collaboration," providing powerful leverage to inventors. For the first time, Quirky allowed inventors to test their ideas in real-time. It, and other sites such as Shapeways, also provided a community for inventors, 3D designers and developers of new 3D printing and manufacturing technologies, permitting these formally disparate groups the chance to communicate, freely exchange ideas, and to work together to solve difficult problems.

The impacts of rapid prototyping will be profound and *some* high-level policymakers such as Al Gore were becoming cognisant of how deep and widespread they could be. (Gore 2013) To highlight how profound and far-reaching the impacts of 3D making will become, consider the example of a man who goes to a toy store to buy a chess set for his grandson's birthday. The traditional approach to manufacturing this chess set was for a designer to produce a collection of standard chess pieces using computer software that was able to generate the injection moulds necessary to mass-produce the pieces. The designs for these injection moulds were then sent to the Chinese manufacturer that would produce a minimum production run that could range from 1,000 to 10,000 chess sets. The sets were then manually packaged and crated before being were loaded into a standard shipping container. This container was transported by truck, or by rail to the coast, where it was then transferred onto a massive container ship that was equipped to carry many thousands of similar containers. After a voyage of about a week, the ship docked at a specialized North American container port, where the container in question was unloaded and stored until a truck could come and pick it up and bring it to a regional distribution facility. At this point the chess set was unloaded from the container and loaded aboard a standard transport truck that brought it to the local toy retailer. The staff unloaded the chess sets from the truck and then eventually put them on display on the store shelves. The customer noticed with dismay that there were only two different types of chess sets for sale, neither of which he felt would impress his grandson. He was therefore forced to make an unsatisfactory purchasing decision based on a highly limited choice of options.

With the new paradigm established by rapid prototyping, the grandfather activated his tablet computer and went to a dedicated website such as iHive3D.com that contained millions of designs for various consumer products. The elderly man selected the *"Board Game"* category and then clicked on the *"Chess Sets"* button, where he found to his delight that there were thousands of different types of chess sets available, designed by some of the finest and most creative artists from around the world. Knowing that his grandson was currently doing a project on Norse mythology, he selected a beautiful ancient Viking style chess set, paid a minimal fee of the few pennies per piece, and then downloaded the designs for each of the different pieces to his tablet computer. He used the 3D design software on his tablet to add some personal touches to each of the pieces, such as engraving his grandson's name on the base of each piece and using the child's likeness to be the head of the king. He then connected his tablet to his personal 3D printer and then proceeded to manufacture each of the individual thirty-two pieces that made up the chess set. At the birthday party, the grandson was delighted to receive such a beautifully-designed, custom-made chess set. (La Puerta 2014)

In this simple example, it was easy to see all of the steps that were made redundant by the introduction of 3D making. With the return of manufacturing to the home after a two hundred year absence, there was no longer any need for expensive injection moulding machines, product packaging, truck or rail transport, containers, container ships, regional distribution facilities, or retail outlets. The magnitude of the disruption to the traditional business model simply boggled the mind. Welcome to your JOOM destiny....

Another example could serve to demonstrate an even greater economic disruption that will be caused by JOOM. Two factors preventing the large-scale adoption of electric vehicles were their high cost and limited range. In March of 2013, it was reported that the company Kor Ecologic produced a vehicle called the URBEE 2, a completely custom-designed car made from only forty 3D-printed, thermoplastic parts. The company produced this design as a first step to see if it would be possible to mass-produce 3D-printed vehicles in the future. (Maxey 2013)

In another example, using off-the-shelf commercial 3D printing technologies, an independent group was able to produce all of the components necessary to build an electric automobile. (Dezeen 2013) Clearly, this approach could be used to mass-produce electric vehicles at greatly reduced prices. While it was obvious that electric vehicles did not need fossil fuels for their operation, it was also true that they were a serious threat to all of "Mister" industries (i.e. Mister Muffler, Mister Transmission, Mister Lube etc.). Electric vehicles were game changers because they did not need cooling systems (i.e. fans, fan belts, radiators, radiator fluids etc.), spark plugs, transmissions, exhaust systems (i.e. mufflers, etc.) and many other components found in standard internal combustion vehicles. Each of these systems represented a major industry that employed thousands of people, industries that would be made redundant with the large scale introduction of affordable electric vehicles. 3D printing will make all of this possible, and much, much, more that was equally disruptive in many different domains. Clearly, 3D making essentially represents the JOOM destiny for many traditional industries.

The types of materials and energy sources that could conceivably be used to print objects in 3D were virtually unlimited. An inventor by the name of Markus Kayser developed a backyard 3D printer that directly used the sun as an energy source and sand as its building material. The focused rays of the sun were used to heat the sand in a small volume to its melting point, leading to the creation of a solid surface. The control software operating on a standard laptop computer allowed the single large lens to track the sun, while another motor positioned the sand processing platform to correctly trace out the solid shape to be manufactured. The object to be printed was first designed as a CAD file, and then divided into a multitude of thin slices. Starting from the bottom of the object, the data to create each slice was sent to the positioning control motors, so that they correctly oriented the volume of sand undergoing melting. After each layer was completed, the operator would manually spread a layer of fresh sand onto the platform, and a subsequent layer of the object would be created. This process continued until the entire object was completed. (Mone 2011)

Additive manufacturing machines were in service within various industries for more than three decades. The problem was always that it required a six or seven-figure outlay to buy one. The big development in the first decade of the 21st century was the rapid collapse of the prices for 3D printers. They first fell to below $100,000, then they dipped to under $10,000 and then dropped to below the critical benchmark price of $1,000. This evolutionary development was so remarkable that a number of high-profile publications found it important enough to report it prominently. For example, Profit magazine announced that the Switzerland-based Geneva Engineering School was marketing a 3D printer kit for a price of about $1,000 US. While innovative, the main drawback of this design was that it was limited to producing proof-of-concept prototypes and scale models of assemblies, rather than actually producing working machines. This remained a big deficiency that had to be addressed by low-cost rapid prototyping systems. (Bright 2012)

In mid 2013, it was announced that a new approach had been developed to create 3D objects. The problem with creating objects layer by layer with a semi-liquid extruded material was always that some sort of structural support had to be provided to keep the material from collapsing under its own weight before it hardened. Various approaches were used to accomplish this, each with its own limitations. Design researcher Brian Harms imagined a way of suspending the molten resin in an inert gel while it was being solidified by exposure to ultra-violet (UV) light. An interesting feature of this approach was that if it was desired to remove a portion of the build material that was accidentally placed in an incorrect location before curing by UV exposure, it could be aspirated back into the nozzle and re-applied in another location.

This approach also overcame another limitation of traditional 3D printing, namely, that each deposited layer had to be given time to harden before the next layer could be applied. Using the supporting gel, the material could be deposited virtually anywhere within the construction volume. This meant that complex assemblies made with multiple materials and requiring a wide variety of different nozzles for their construction could be handled without difficulty. It could be estimated that the development of new gels and supporting materials would allow assemblies to be printed in complex combinations of plastics, metals and other exotic materials. This meant that rather than being limited by the fabrication technology, future designs would only be limited by the creativity of the designers themselves. (Teschler 2013)

Like any technology, 3D making could be, and in fact already had been, used for malicious purposes. The press provided extensive coverage of home-based entrepreneurs who used their home-based 3D printers to make various weapons from scratch such as pistols and fully automatic weapons. In another report, a discrete device that had been produced with a 3D printer was attached to an ATM and was used to steal data from debit cards and to record PINs as they were typed into the ATM by unsuspecting customers. (Krebs 2011)

As upsetting as this was to the average person on the street, the fact was that *any* technology could be used for evil – the laser used to save a person's vision by welding on a new cornea, could equally be used to blind. It was not the technology that was to blame, but the people using it. In coming years, it is hoped that the benefits that will accrue to humanity from the proliferation of 3D technologies of all kinds will gradually push the malicious applications to the back of the public consciousness.

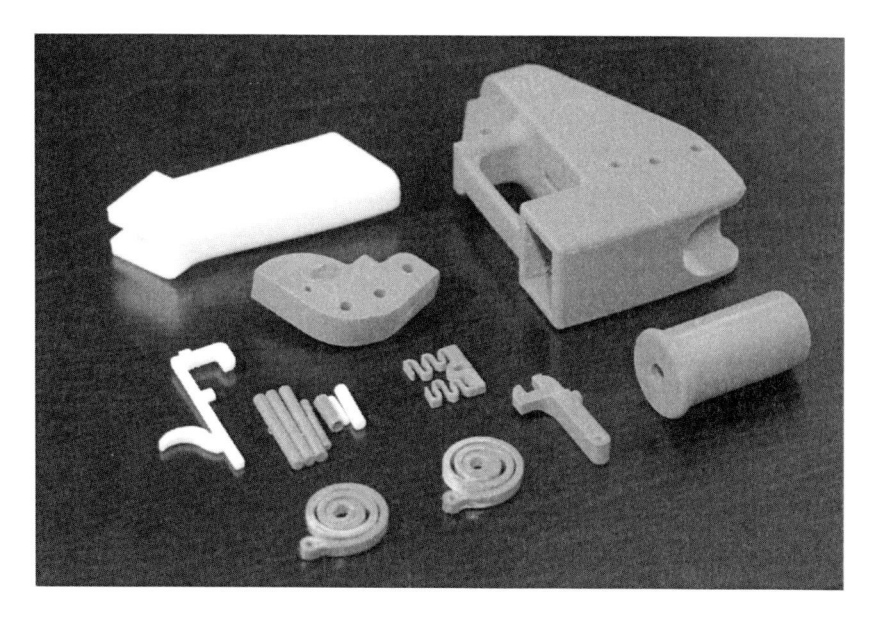

Figure 2-7: The 3D printed components used to build the Liberator hand gun. Source: Vvzvlad

Just-On-Order will not only impact the way manufacturing is done but will also greatly affect business processes. For example, the Canadian company D-Wave Systems began to sell its revolutionary quantum computers to large industrial concerns such as Lockheed Martin in 2011. A unique feature of D-Wave was that it offered a developer portal that allowed potential users to find out everything that they needed to know about the operational details of this next generation computer, and to learn for themselves how they could apply it to solve their specific business or technical problems.

Solutions where and when you need them – this is the essence of your JOOM destiny.

Just On Order Electronics

A company named New Energy Technologies was one of the first to commercialize "spray-on" solar cells only a fraction of millimetre thick that were capable of converting any surface into a photo-voltaic (PV) energy generating cell. The potential of this development was virtually unlimited since every major city on earth was full of high-rise buildings that could have each of their windows converted into an energy-generating surface. The first generation of energy-generating films were created by using a high pressure spray nozzle operating an oxygen-free environment to assure that no airborne particles contaminated the surface. Since this process was rather expensive, it was highly desirable to use a specially-adapted 3D printer to lay down the solar cell material, in liquid form, to create a precise energy-generating layer. Since this layer could be made to adhere to a wide variety of substrates including glass, steel, plastic and wood, the potential existed to convert virtually every surface in a modern office building into an invisible PV cell generating energy for the occupants. (Snieckus 2010)

One of the really exciting frontiers of 3D making was its application to the manufacturing of very small devices, on the scale of billionths of a metre (nano-metre). There were two general strategies for building nano-scale devices. The first was to shrink larger devices to ever smaller sizes (the top-down approach), while the second approach was to assemble atoms one by one into larger structures (the bottom-up method). A number of researchers have sought various biological agents that could be harnessed to build functioning machines by moving atoms and molecules into specific predetermined positions. But what biological agent could possibly be able to move individual atoms around?

In the closing years of the 20th century, researchers began to examine the properties of a virus known as M13 that attacked bacteria but was harmless to human beings. Using a process known as Directed Evolution, this virus had the property of being able to bind to specific organic compounds, metal ions, and semiconductors. Once a virus with the specific desired features was found, using a process known as amplification, trillions of copies of it could be manufactured and they could be used to assemble a tiny structure, initially molecule by molecule, and then ultimately in the future, atom by atom. A virus that was designed to link to gold ions could be used to construct a gold wire several centimetres long that could then be woven into a gold fabric. A variant of this virus could be used to construct a gold film that was several centimetres square, but only a micron thick. (Ross 2006)

Variants of the M13 virus were designed to construct the electrodes used in new generations of ultra-light lithium-ion batteries. Others were designed to bind to cobalt oxide while others were capable of storing lithium ions. It was then possible to create a thin lithium electrolyte film that was used in a storage battery that could be moulded to fit into virtually any available space. The viruses could be used to discriminate between the semi-conductors gallium arsenide and gallium nitride, which gave them the ability to construct perfect semi-conductor chips with properties superior to chips constructed using traditional methods. In an unexpected medical application, the viruses were found to be capable of binding to specific molecules found uniquely in cancer cells, making them capable of acting as living detectors of specific diseases.

In another ground-breaking technological innovation, researchers discovered that it was possible to spray-paint a fully-functional lithium-ion battery onto virtually any surface. The challenge was to find a way of converting all of the five major components of a battery, including the current collectors, cathode, anode, and the polymer separator into paintable form. This capacity to create Just-On-Order-Batteries was a significant advance for all forms of alternate energy because it meant, for example, that ceramic tiles on the outside of a house could all be converted into very inexpensive paint-on batteries that could easily be connected to thin solar cells, creating a cheap energy collection and storage system. (Lamb 2012, Dodson 2013)

Just-On-Order-Molecules - Nano JOOM

IBM research developed a micro-3D printer that had a nozzle so small that that it could create patterns on polymer surfaces as small as 0.01

mm (1×10^{-5} m) in size. As notable as this accomplishment was, this was only considered to be the first step towards the long-term goal of developing nano-3D printers capable of working at scales approaching 0.000001 mm (1×10^{-9} m). (Shingler 2014) In fact, researchers managed to develop a 3D printing technique that succeeded in stacking nano-scale polymer threads on top of each other to create a structure, but the type of shapes that could be created by this method was limited. (Nature 2014)

The ultimate expression of JOOM was in the creation of atomic-scale assembly lines based the processes used by living cells. It was possible in principle to reproduce all of the elements of the modern assembly line at a scale of the order of a billionth of a metre. The central tool that was used for nano-scale engineering was the DNA molecule, which has long been known to have the property that it would assemble into specific shapes depending on its well-understood chemical properties.

How does one build a molecular scale robot (*nanobot*)? Building at the scale of one billionth of a metre, it was possible to start with a circular, single-strand DNA molecule that was then mixed with many short pieces of complementary DNA. By carefully selecting which complementary DNA sequences were used, the device self-assembled into a predetermined 3D structure. According to Liedl et al. (2010):

> *Double helices fold up into larger, rigid linear struts that connect by intervening single-stranded DNA. These single strands of DNA pull the struts up into a 3D form -- much like tethers pull tent poles up to form a tent. The structure's strength and stability result from the way it distributes and balances the counteracting forces of tension and compression. This architectural principle -- known as Tensegrity -- has been the focus of artists and architects for many years, but it also exists throughout nature. In the human body, for example, bones serve as compression struts, with muscles, tendons and ligaments acting as tension bearers that enable us to stand*

up against gravity. The same principle governs how cells control their shape at the micro-scale.

Exactly how was it possible to make a nanobot "walk" on a molecular scale? Research showed that it was possible to make a rudimentary nanobot with three or more legs made of enzymes consisting of different DNA sequences. The nanobot could "walk" by binding one of the front legs to an "anchor" on a molecular track and then cutting a DNA strand on one of the back legs, before using molecular forces to advance the leg to a new forward position. Essentially the nanobots were simply following the chemical clues that were programmed into the specific path that they were walking.

What could such a nanobot be good for? By building a programmable DNA path along with a fleet of independently controlled nanobots that were designed to be miniature "pickup trucks" carrying payloads consisting of parts to build additional nanobots, the result was a miniature assembly line operating as a nano-scale factory.

In another experiment, researchers at NY University used DNA based nanobots controlled by chemical signals to assemble gold particles in more than half a dozen ways. In contrast to existing nanotechnologies, this was considered to be a significant breakthrough because these programmable nanobots were highly suitable for a multitude of human medical applications because the DNA that was used to build them was both biocompatible and biodegradable.

This meant that it would soon be possible to build nano-scale medical devices and drug delivery systems based on synthetic virus mimics that could be used to introduce drugs directly into diseased cells. A nanobox could be constructed with a chemical "lock" that would only spring open in response to a specific signal and would ensure that drugs were only released into the body when and where desired.

Nanobots could be programmed to guide human stem cells into a regeneration mode that would allow them to restore injured organs. Since stem cells responded differently depending on the inter-cellular forces and chemical gradients around them, nanobots could be used to create the right environments around specific groups of stem cells that would lead to the generation of specific types of cells such as, for example, bone, liver or heart cells. This would permit Just-On-Order-Tissue engineering at the molecular scale to finally become a reality.

JOOM and Animal Compassion

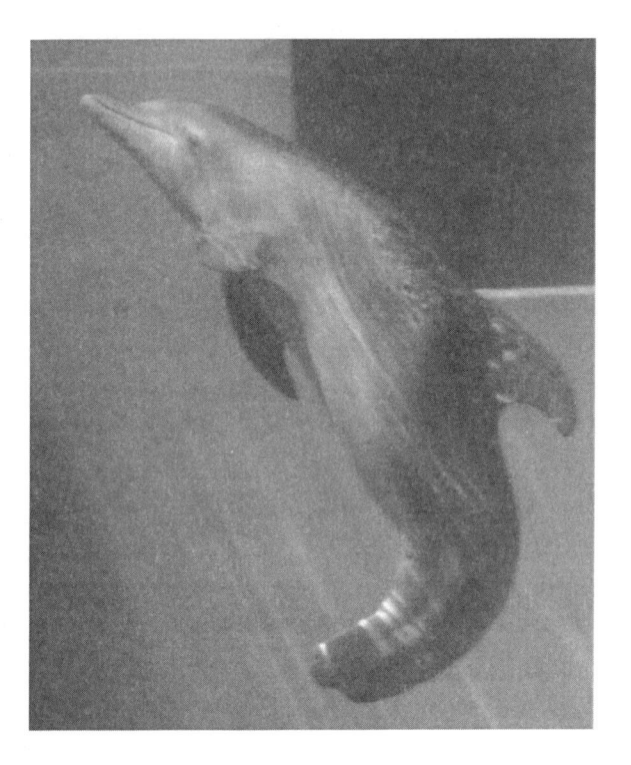

Figure 2-8: Winter the dolphin shortly after
losing her tail fin. Source: Paul

At the age of only three months, Winter the dolphin entangled her tail fin in a crab trap. Her injuries were so severe that she ending up losing this vital part of her anatomy. Since a tail fin was absolutely necessary for marine mammals to lead a normal life, the humans at the Clearwater Marine Aquarium who had taken Winter into their care decided to use leading edge technology to create a new tail for the unfortunate little dolphin. The first step was to use a scanner to create a 3D image of the animal's remaining lower body and then to use CAD software to fill in the missing tail. Then a thermoplastic elastomer was shaped to tightly fit over the tail stub, and this was in turn attached to a plastic swimming fin. This approach proved so successful, that other organizations such as the Walter Reed Army Hospital began using it for their human patients. An enhancement that greatly speeded up the process of creating the prosthetics was to use a 3D printer to build both the elastomer sleeve and the ABS plastic fin directly from the scanner data.

In a similar story, Molly the pony lost the lower part of her leg to a vicious pit bull attack. Her first prosthetic was assembled out of acrylic, aluminum and fibreglass, and over the years, the vet team at Louisiana State University continued to use Molly's experiences with the artificial leg to construct increasingly capable models. The ability of 3D printing to made accurate 3D constructions in a variety of materials made it ideal for applications like making custom designed artificial limbs for virtually any living creature that required one.

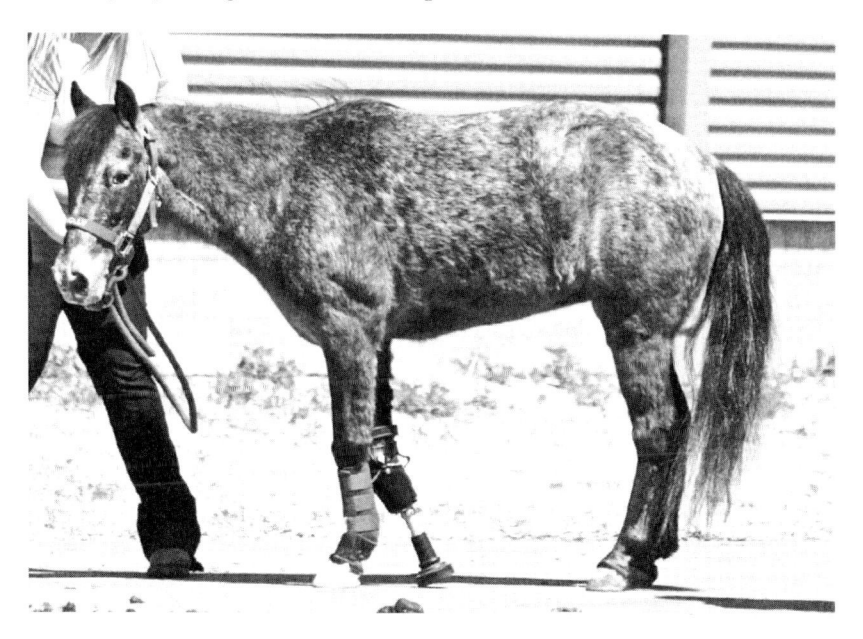

Figure 2-9: Molly the pony standing confidently
on her 3D printed lower leg. Source: Jean

Bio-JOOM – Biomimicry

"Nature to be commanded must be obeyed." Anonymous

Anyone who took the time to examine the bones of a bird skeleton observed that far from being sold cylinders of calcium, the bones were made up a hollow lattice of thin calcium fibres that created a structure that was at the same time incredibly strong and exceptionally light.

Using 3D making technologies, it was now possible to economically mimic this natural approach to maximizing structural strength by creating metal cylinders for products such as bicycles and tents as see-through lattice structures instead of traditional hollow metal tubing. The new structural material was known as IsoTruss and research showed it to be far stronger than tubes made of traditional materials such as aluminum or steel. When 3D printers are eventually developed that can work with carbon fibre, the material will be even stronger and will be a significant game-changer in the world of advanced composite materials. (Dyer 2008, Lawn Rider 2013)

MIT's world famous Media Lab had a less well known sub-group known as the Mediated Matter Group. According to the group's website:

> *"The Mediated Matter Group focuses on biologically inspired design fabrication tools and technologies aiming to enhance the relation between natural and man-made environments. Our research field entitled Material Ecology integrates computational form-finding strategies with biologically inspired fabrication. This design approach enables the mediation between objects and environment; between humans and objects; and between humans and environment. Our goal is to enhance the relation between natural and man-made environments by achieving high degrees of design customization and versatility, environmental performance integration and material efficiency. We seek to establish new forms of*

> *design and novel processes of material practice at the intersection of computer science, material engineering, design and ecology, with broad applications across multiple scales." (Media Lab 2013)*

Fundamentally speaking, a creature like a silkworm was nothing more than a living 3D printer. Its function was to take raw material (silk) and to eject it at a controlled rate from a nozzle (its abdomen) to form a pattern determined by a pre-established software program operated by a microprocessor (its brain). Early experiments discovered that the spinning patterns of silkworms could be influenced by modifying the environment they were living in. After many years of study, sufficient data was gathered to write a program that could efficiently direct the silk output of thousands of silkworms so that their output could be used to construct structures useful for human applications.

The first step was for a CNC machine to construct a metal scaffold that was used to host a population of silkworms. A *"seeding"* pattern of silk fibres was then applied to this scaffold that induced the worms to fill in the gaps in the network. In one experiment that explored the ability of silkworms to make architectural structures, the result was a beautiful building known as the Silk Pavilion. In the words of Prof. Neri Oxman:

> *"The Silk Pavilion explores the relationship between digital and biological fabrication on product and architectural scales. The primary structure was created of 26 polygonal panels made of silk threads laid down by a CNC machine. Inspired by the silkworm's ability to generate a 3D cocoon out of a single multi-property silk thread (1 km in length), the overall geometry of the pavilion was created using an algorithm that assigns a single continuous thread across patches providing various degrees of density. Overall density variation was informed by the silkworm itself deployed as a biological printer in the creation of a secondary structure. A swarm of 6,500 silkworms was positioned at the bottom rim of*

the scaffold spinning flat non-woven silk patches as they locally reinforced the gaps across CNC-deposited silk fibres....Affected by spatial and environmental conditions including geometrical density as well as variation in natural light and heat, the silkworms were found to migrate to darker and denser areas. Desired light effects informed variations in material organization across the surface area of the structure. A season-specific sun path diagram mapping solar trajectories in space dictated the location, size and density of apertures within the structure in order to lock-in rays of natural light entering the pavilion from South and East elevations. The central oculus is located against the East elevation and may be used as a sun-clock. Parallel basic research explored the use of silkworms as entities that can "compute" material organization based on external performance criteria. Specifically, we explored the formation of non-woven fibre structures generated by the silkworms as a computational schema for determining shape and material optimization of fibre-based surface structures." (Oxman 2013)

Just-On-Order-Machines

The Defence Advanced Research Projects Agency (DARPA) long sought a Common Operating System (COS) that could be used equally well in all types of autonomous robotic machines including reconnaissance drones, helicopters, pack-mules, supply trucks and assembly robots. The successful development of a COS would lead to an explosion in the kinds, numbers and applications of autonomous robots.

It was also proposed that the development of open-source robot blueprints could dramatically cut the cost of robots due to new open-source hardware-sharing systems. Similar to open-source software for computers, this new robot-development platform would allow participants to share their designs so that other developers could adapt or improve on them. The hope was that by sharing hardware and software development across thousands or even millions of developers, costs would plummet while creativity exploded. The effectiveness of this approach could be demonstrated by considering a typical nursing home care-giving robot that previously cost more than $350,000. Open-source development of the next generation of this type of robot combined with use of a robotic COS could lower costs for this system to under $25,000. This approach clearly results in significant savings for cash strapped medical systems everywhere.

The next goal was to reduce the cost of fully-autonomous (FA) robots even further, hopefully to below $1,000. This could be accomplished when such machines acquired the capability of manufacturing new generations of FA machines. A number of 3D printers were endowed by their creators with the capability of reproducing themselves, such as the open source RepRap project (www.reprap.org) and Makerbot (www.makerbot.com). While these models led the way, many new machines will come to market that will also have this capability. Since the majority of the critical parts needed by a typical rapid prototyping machine were made of plastic, these machines were fully capable of manufacturing the parts that essentially allowed them to create copies of themselves. All the user had to provide were the electronics. But the pace of innovation in the 3D making field was so rapid, that it will soon be possible to print the electronics as well, leading to the creation of machines with full self-reproducibility.

What was certain was that the whole field would explode with the development of low-cost 3D fabrication machines with the capability of rapidly producing structurally complete metal parts that could actually be used to replace traditionally manufactured spare parts in real-life operational environments. When this happened, all of the limitations currently holding back the full deployment of 3D printers at all levels of society would be removed and the full impacts of the revolution, for better or worse, would finally be felt by all.

Dr. Hod Lipson, a computer science professor at Cornell University and his graduate student, Michael Schmidt, developed a computer program known as Eurega that was able to take data generated by a physical system, such as a pendulum, and from an analysis of the motion, derive the fundamental physical laws describing the operation of the system. The program was able to generate new discoveries in a wide variety of fields including biology and biophysics. The program made an analysis of the metabolic dynamics of a bacterial cell and succeeded in generating an equation that was highly elegant and that described all of the known biochemical operations within the cell. The difficulty was that no human being could understand how or why the equation worked or what new scientific principles it was demonstrating. (Manjoo 2011)

In a number of highly technical areas such as Computational Fluid Dynamics (CFD), a number of programs were developed that were able to autonomously write software capable of simulating many types of complex fluid-flow systems. While these machine-generated codes produced results that closely matched the physical systems that they were modelling, no human programmer could decipher or "de-bug" these codes, nor could they unravel the flow of logic within the artificially generated computer code. What were the implications when even the most intelligent humans were no longer capable of fully understanding the technologies they were forced to depend on? (Bentley Systems 2009)

The observation that could be made about these developments was that scientists and engineers had entered a brave new world where machines were doing things that humans could not currently understand and perhaps never would be able to understand. This could mean that machines would soon be the primary generators of new Just-On-Order-Knowledge and that humans will just have to hang on and go along for the ride. The full impacts and long-term consequences of this remained to be fully explored.

Just-On-Order-Microbial Products

It was long known that certain bacteria could be used to extract metals from solutions. For example, the bacteria *Cupriavidus metallidurans* was used to extract pure gold from a toxic solution of gold chloride. During this process, the bacteria were "fed" by exposing them to a hydrogen saturated atmosphere. This method was inefficient however, and so its main use was to clean up metals from various wastewaters before their discharge to the environment. Many other species of bacteria were put to work to solve various production problems that would have been uneconomical using any other process. *Escherichia coli* was harnessed in the process for manufacturing pristinamycin, an antibiotic used to fight staph infections. *Saccharomyces cerevisiae*, a species of yeast, was adapted to the manufacture of synthetic anti-malaria drugs. *Methanobacterium palustre* was encouraged, with the aid of an electric current, to produce pure methane. The promise of fracking was of a new supply of natural gas, capable of meeting North American demand for more than two hundred years and that was not subject to the uncertainties of Middle Eastern political, ethnic and religious tensions. All that had to be done was to find new ways of dealing with the significant adverse environmental impacts of this new technology. Use of the specialized bacteria, on the other hand, held the promise of virtually unlimited, sustainable and environmentally safe production of methane forever. (Mone 2013)

The long-term trends indicated that it will eventually be possible to build molecular scale 3D printers that will be able to assemble first viral, and later bacterial genomes from scratch, base-pair by base-pair. Once a detailed understanding was gained of the exact purpose of each of the billion or more base-pairs in a typical bacterial genome, it will be possible to design and construct living bacteria with specific capabilities not found in any natural organisms. The incredible potential this capability will offer to humanity to improve the world can only be the subject of speculation at the time of this writing. But it is also certain that there will also be limitless potential for harm, in the form of, for example, new bacterial warfare agents and other particularly nasty applications that cannot be foreseen at the present time.

Looking further into the future, once the exact functions of various groups of base pairs have been determined and recorded into a data base, it will be possible to use molecular 3D printers to construct entire genomes from scratch to generate Just-On-Order-Life Forms. These could be exact copies of organisms that lived in the past, such as Apatosaurus, or could be life forms that previously existed only in the imagination of a systems biologist.

Jet Engines Produced by 3D-Printing

In May of 2013, the CEO of General Electric (GE), Jeffrey Immelt, announced that it would be a goal of his company to continuously increase the number of components of GE's leading edge jet engines that were produced by 3D making, dramatically reducing waste and manufacturing times.

A small company located in Cincinnati, Ohio, named Morris Technologies invested a large amount of venture capital into a number of leading edge 3D printing machines using a variety of different technologies. Their goal was to eventually use this technology to build as many of the functional parts of a modern jet engine as possible. However, before they could market the parts made by their process, GE moved in and purchased the entire company, saving themselves years of tedious and expensive research and development work.

It was later reported that GE was using the technology obtained from Morris Tech. (as well as technology obtained from Rapid Quality Manufacturing) to produce a new engine (known as LEAP) that used blades made of ceramic composite materials as well as a number of functional parts made with 3D printers. It was claimed that eventually, not only would the blades be made by rapid prototyping, but virtually the entire engine would be manufactured using this new technique. (Bullis 2013)

The Factory of the Future

Companies such as DaimierChrysler spoke of the need to develop the "Factory of the Future" (FF). (DaimierChrysler 2006) The first step in the FF concept was to develop a digital representation of all aspects of a particular facility including the production facilities, machines, robots, individual assemblies, products, ergonomic tests, etc. These representations included 3D designs for each item, system and sub-system. The great advantage with this approach was that before a single shovelful of earth was turned for a new plant, the overall reliability of every system and plant process as well the quality of every future manufactured product that would be coming off the future assembly line could be verified to a high degree of accuracy beforehand. As good as this sounded and as advanced as the FF concept was, it was made virtually obsolete with the introduction of the Advanced JOOM Process (AJP).

Assembly lines, whether manned by humans or increasingly, by robots, were a fundamental fixture of manufacturing since the time of Henry Ford. AJP did away with all of the features that had traditionally been associated with heavy manufacturing. First of all, sales and marketing departments were no longer needed as the company presented its brand and desired public image of itself through an interactive web site. Projections no longer had to be made of future sales of particular products, because customers actually told you on an hour-by-hour basis what products were wanted and in what quantities. In other words, no product was manufactured until there was a firm, paid order in the data base, so instantly, all of the costs associated with sales projections and inventory disappeared.

If the customer wished to order a car, they would first go to the company web site and answer questions relating to the desired design such as whether they wanted a two door, four-door, or a hatch-back model, number of cup holders, exterior and interior colour preferences, four or six cylinder engine options, diesel, gas or electric powered etc. In other words, the entire car could be custom designed by the client. After the entire vehicle design had been specified, the customer would then be brought to the financing page where, either a lease-to-own, business lease, or an outright purchase was set up.

Once all of this was arranged, when the customer pressed the "*Manufacture*" button, all of the relevant information was sent to the computers located in the "Assembly Sphere," an area no bigger than the bathroom in a typical monster home, where the entire vehicle was assembled in one location, surrounded by dedicated 3D printers, some spitting out metal parts, others printing composite structural plastic and still others making the electronics and the transparent polycarbonate for the windows. The raw materials fed to these rapid prototyping machines consisted of sintered metal, plastic granules, glass and metal fibres. In less than an hour, the additive manufacturing processes produced a complete vehicle, exactly as designed by the customer and ready for delivery. The process was so efficient, producing virtually no waste or unnecessary production, that a perfect custom designed vehicle could be delivered to the customer's door within two business days, for a price that ranged from one-half to one-tenth of that charged by the major car manufacturers for a mass produced vehicle built on a traditional assembly line. There was absolutely no way that any organization that continued to use assembly lines to manufacture stuff could have any hope of competing in the long-term.

The entire process was automated, with the only personnel present in the factory being a plant supervisor (an engineer), an assistant, a software engineer and a robotics/3D printing technician who kept the whole system well maintained and functioning smoothly. There would be no need for the thousands of salaried workers that were normally found in traditional manufacturing organizations. Another example of your JOOM Destiny...

—

The construction industry employed millions of salaried employees at above average wages. Construction methods and technologies had changed little in more than half a century. The industry was about to experience what could be termed a "*workquake*." The Office of Naval

Research and the National Science Foundation investigated the possibility of using field deployed 3D printers to construct entire buildings. Concrete printers could produce an almost-ready to occupy structure in less than a day, at a cost that was only a fraction of the cost a conventional project. In addition, it was found that the walls produced by 3D printers were almost fifty percent stronger than traditionally poured walls. (Bushey 2014)

There were many profound implications associated with these developments not the least of which was that 3D printers reduced the need for human resources in the construction industry by over ninety percent. Until recently, the construction industry was one of the last areas that had resisted automation. JOOM was about to change this dramatically.

—

Solar PV technologies played a central role in the national alternate energy policies of many nations around the world. The main contributor to the high costs of these installations was the difficulty in producing highly purified polysilicon, the main raw material for the cells. Continued cost reductions depended on driving the economics of scale by increasing manufacturing plant capacities to above 1 GW per year, decreasing wafer thickness to below 120µm (about the thickness of a human hair), increasing processing speeds, augmenting yields by the development of improved methodologies such as manufacturing execution systems and automatic/statistical process controls, reducing the areas shaded by contact wires, application of advanced anti-reflection coatings and by other improvements in the cell design.

All of these innovations contributed to driving cell efficiencies to beyond twenty percent. Further advances will require the use of more radical technological innovations. Significant improvements will occur with the introduction of molecular 3D printers that will be able to design ultra-thin film PV panels atom by atom, resulting in perfect crystalline structures that offered high levels of incident photon absorption, high conversion efficiencies and ultra-low resistance to electron flow.

In the future, the introduction of new PV technologies, such as Quantum-Dot PV panels for example, may eventually push conversion efficiencies to beyond eighty percent. Each quantum-dot can be tuned to absorb a specific photon frequency, so billions of the dots would be able to absorb and convert virtually all of the incident solar radiation striking a panel into electrical power. The widespread availability of such panels would mean that each house could be converted from being power consumer, to being a net power producer. The utility power grid would then only exist not to supply power, but to redistribute it. In many jurisdictions, power companies would be forced to pay homeowners for the power they injected into the grid, rather than the other way around. The economic, political and social implications of such a development would be significant.

The Limitations of JOOM

As with all revolutions, it will take time to iron out all of the problems and to overcome the limitations encountered by the technologies associated with JOOM 1.0. At the time of this writing, a number of industry observers listed numerous reasons why traditional industries did not yet have much to fear from the so-called 3D Revolution. (Johnson 2013)

According to Johnson (2013) and others, 3D printing will not be viable for the foreseeable future. A generation ago, "experts" asserted with confidence that computers would remain multi-million dollar behemoths that would only serve the needs of business and industry and that they would remain too expensive for ordinary individuals to own. And besides, people had no need for computers in their homes because what on earth would they do with them? The rapid and relentless forward march of technology (as evidenced by the constancy of Moore's Law for more than fifty years) strongly supported the view that Johnson and other critics suffered from a serious lack of vision regarding additive manufacturing and supporting technologies. Here then is a possible conversation between a critic of 3D printing and a visionary who could see beyond the short-term limitations and hype, imagining what the technology could eventually become.

Critic: "*3D printers are too difficult for ordinary individuals to ever use. Only technical geeks and hackers have the skills and patience to make 3D printers work and it requires huge amounts of ongoing maintenance to keep them working.*"

Visionary: "Early cars, sewing machines and personal computers were touchy, difficult to use and were not user friendly. However, technical visionaries soon smoothed out the rough edges and made these and hundreds of other technical consumer products much easier to use. Just as the advent of the USB standard made printers and other computer peripherals into true "plug-and-play" devices, new technological advances will greatly increase the usability and maintainability of 3D printers and supporting technologies. Also, these innovations will happen over a time span of five and not fifty years, as claimed by those with limited technical vision."

Critic: "*There are not many things to print. There is not a whole lot of interesting stuff out there that can be downloaded and printed meaning that users have to design for themselves what they want to print using highly difficult to use and expensive professional 3D design software packages such as AutoCAD, ProEngineer and Solid Works.*"

Visionary: "In the early days of automobiles, there were not very many miles of paved roads to drive on, not many places to get gas and virtually nowhere to get help after a breakdown. That all changed rapidly as companies such as the Ford Motor Company made the automobile into an affordable consumer item, the federal government undertook the construction of the national highway network, and the oil companies built networks of service stations. Highly user-friendly web sites such as **iHive3D.com** will soon be hosting the designs for millions of useful items in thousands of categories that users can download for only pennies apiece, cutting the dependence that individuals previously had on central manufacturing facilities to make the consumer goods. This will bring manufacturing back into the home, just as it was before the beginning of the industrial revolution."

Critic: "*The stuff that you can print is not really useful. Sites like Thingiverse are full of toys, art and fanciful items but ultimately useless gadgets that have little practical use and these can only be printed in plastic that has little structural strength and so can break easily. Plus it is not possible to make any components with internal moving parts, so each part has to be created separately and then all of the parts have to be laboriously assembled to create the final product.*"

Visionary: "New Web sites have begun appearing containing highly useful and otherwise hard to obtain items such as designs for hard-to-get spare parts for devices that are no longer being produced such as vintage cars, steam engines and typewriters. A chess player going to a retail store seeking a set of pieces will typically have a choice between only two or three different styles. In contrast, the same individual shopping on a 3D design Web site such as **iHive3D.com**, will eventually be confronted with a selection of thousands of different custom made chess pieces that can be downloaded for pennies apiece. Just as modern laser printers can now print in any color using any font, the next generations of personal 3D printers will be able to print virtually anything in essentially any material including stainless steel, copper, carbon fibre and gold. Contrary to the early 3D printed parts that had no structural strength and that were suitable only for the production of non-functioning, shelf-sitting prototypes, the new printers will produce fully functional parts with the same (or even greater) structural strength and wear resistance as the original machined parts that they are replacing. The newest 3D printers are so accurate that they are capable of printing an entire automobile transmission with its dozens of internal gears and moving parts in one go, without the requirement for any post manufacturing assembly. The time savings potentially offered by this approach are enormous."

Critic: *"Each of the major CAD packages use proprietary formats that are completely incompatible with each other. Multiple industry formats has led to increased project complexity where engineers and designers working in different locations and using different platforms are unable to exchange designs with each other. The use of poor data translation programs results in numerous downstream problems, and in many cases, data imperfections only become apparent when attempts are made to manufacture the components. Additional difficulties arise when new CAD formats are introduced, making it difficult or impossible to read legacy files."*

Visionary: "Just as human languages have gradually became transparent to each other because of the development of increasingly sophisticated computer translation programs, 3D CAD designs have also became more inter-accessible to each other. A software package such as 3D InterOp Suite provides a highly extendable, modular, "plug-n-play" translation tool. It will accept designs made in a variety of different platforms including CATIA Vx, Solid Works, Pro/Engineer, Solid Edge, Inventor, IGES and many others. It extracts the geometry, topology and graphical parameters, then it automatically repairs these where necessary and adjusts the tolerances, and then outputs the data into a specified format including Product Manufacturing Information (PMI) or STL formats that are vital if the product is to be accurately rendered on a 3D printer or manufacturing facility. The US military is working on the development of a universal standard for 3D rendering that could eventually be adopted by the private sector."

Critic: *"There are huge legal barriers to personal manufacturing. Large multinational corporations will not tolerate having their Intellectual Property (IP) stolen by individuals engaged in home manufacturing and they will assert their control by suing everyone engaged in the scanning, replication and production of proprietary objects."*

Visionary: "Large enterprises, organizations and manufacturing concerns seem to think that they have cornered the market on creative genius and innovation. In fact, the majority of the most creative individuals do not work for large organizations and are often to be found tinkering away in dusty attics or grimy basements. When given the proper empowering technologies and the means to propagate their work, these individuals are perfectly capable of innovating on their own, making their own intellectual property and introducing their own product/service innovations. Witness the development of the Linux PC Operating System that was created by the efforts of thousands of unpaid volunteer software developers from around the world. Web sites hosting 3D designs such as **iHive3D.com** will eventually be filled with hundreds of millions of original creations made by millions of innovative designers from around the world that are free from any copyright or IP entanglements. Many of these products will be superior to anything offered by the large, established, traditional manufacturers."

Yes, it was true that the legal departments of the giant music, TV and movie industries went absolutely berserk when technology evolved to the point where it allowed individuals to easily copy proprietary, music, media, artistic and cultural products without licence and without the payment of any royalties to the owners of copyright. Eventually a fair compromise was reached where sites such as iTunes and NetFlix were able to sell consumers proprietary media products at a fair price, while allowing the artists and the producers to maintain fair profits as well. All this was made possible by the advent of cheap MP3 players and tablet computers along with the ability to easily conduct credit card transactions online. The same thing will happen in the additive manufacturing sector where large manufacturing concerns will upload their designs for thousands of spare parts for each of the products that they produce, and consumers will then happily pay a nominal fee to download these designs so that they can create these parts at home on their personal 3D printers. Quality standards will be maintained by the software built into the printers themselves, so that substandard, structurally inferior parts will simply not be capable of being produced.

Conclusion

The world is now in the grip of a new industrial revolution that will result in the complete digital transformation of manufacturing and in its linkage to web-based services and media. This transformation will represent a fundamental paradigm shift in the way all items are manufactured because it will eventually mean that anyone will be able to make virtually anything, anytime, anywhere. There will no longer be any economics of scale and there will be the promise of universal customization.

The new technologies have eliminated most of the barriers that used to restrict individual, and small-scale manufacturing, invention and innovation. Virtually any idea can now be transformed into the reality of an actual physical product. Using social media and other internet technologies, these innovations can then be easily distributed throughout the world at the touch of a button.

When a complete life-cycle analysis was conducted on a variety of 3D printed objects, it was found that when compared to traditional manufacturing processes that had to ship their products throughout North America, printed objects required from about forty to sixty-four percent less energy. The savings came from using less raw material and by the elimination of shipping costs. When printed items were made from polylactic acid (PLA) which was made from renewable resources such as cornstarch instead of fossil fuels, 3D printed objects not only required less energy to build, but they also produced significantly less CO_2 emissions, making them much greener than their traditionally manufactured counterparts. The 3D printed objects also cost significantly less to create than the manufactured ones, sometimes up to thousands of dollars less. The greatest savings were achieved for speciality, custom-made items such as orthotics or dentures. (Wittbrodt et al. 2013)

The costs of 3D printers and scanners are falling dramatically. For example, only a short time ago, an entry level industrial scanner cost more than $20,000. Then scanners with virtually equal capabilities became generally available for about $2,000. At the time of this writing, a hand-held scanner has just been released with most of the capabilities that the home hobbyist could want for less than $400. The general collapse of the prices for all types of 3D technologies is expected to continue.

It can be concluded that the new business paradigm, centered on 3D printing and supporting technologies, offers the possibility of making individually customized objects at lower cost, using less energy and with the generation of significantly less waste. It will be extremely hard for traditional manufacturers to overcome these overwhelming advantages in the long-term.

Web Resources

Potential of 3D Printing: http://www.youtube.com/watch?v=LRv4jp-hhBE

3D Printer in Action: http://www.youtube.com/watch?v=MwuGbnjKJBc

Sample 3D Print: http://www.youtube.com/watch?v=vKDVVb2blo0

3D Metal Printing: http://www.youtube.com/watch?v=i6Px6RSL9Ac

3D Printed Metal Gun: http://www.youtube.com/watch?v=5ml_V7VyOuw

3D Printing of Houses: http://www.youtube.com/watch?v=ehnzfGP6sq4

3D Printing of Buildings:
http://defensetech.org/2014/01/20/navy-helps-fund-3d-printing-of-buildings/

Printing Replacement Car Parts:
http://3dprintingindustry.com/2014/02/02/3d-printing-replacement-car-parts/

Photo Credits

Figure 2-1: http://dayton.hq.nasa.gov/IMAGES/LARGE/GPN-2000-001926.jpg

Figure 2-2:
http://upload.wikimedia.org/wikipedia/commons/5/55/
Geely_assembly_line_in_Beilun%2C_Ningbo.JPG

Figure 2-3:
http://commons.wikimedia.org/wiki/File:Sharp_3_
Axis_Vertical_Mill_CNC_Converted.jpg

Figure 2-4:
http://upload.wikimedia.org/wikipedia/commons/
thumb/8/84/St%C3%A9r%C3%A9olithographie.
jpg/1119px-St%C3%A9r%C3%A9olithographie.jpg

Figure 2-5:
http://upload.wikimedia.org/wikipedia/commons/thumb/c/c8/
Imprimante_3D_-_cit%C3%A9_des_sciences_-_Fab_Lab.JPG/1280px-
Imprimante_3D_-_cit%C3%A9_des_sciences_-_Fab_Lab.JPG

Figure 2-6:
http://upload.wikimedia.org/wikipedia/commons/d/
df/Rally_Fighter_Local_Motors_2.jpg

Figure 2-7: http://commons.wikimedia.org/wiki/File:Liberator.3d.gun.vv.02.jpg

Figure 2-8;
http://commons.wikimedia.org/wiki/File:Winter_tail-
less_bottlenose_dolphin_croped.jpg

Figure 2-9: http://commons.wikimedia.org/wiki/File:Molly_the_Pony.jpg

References

Advanced Manufacturing (2005) "New Rapid Prototyping Process Catches Hold," January/February, p.12

Bentley Systems (2009) "Bentley Ships Green Building Design Series for US, Canada," *TenLinks.com*, March 17, http://www.ten-links.com/news/PR/BENTLEY/03170609_green_series.htm

Bright, M. (2012) "Mass Production Goes DIY," *Profit Magazine*, February

Bullis, K. (2013) "A More Efficient Jet Engine is Made from Lighter Parts, Some 3D Printed," *MIT Technology Review*, May 14. http://www.technologyreview.com/news/514656/a-more-efficient-jet-engine-is-made-from-lighter-parts-some-3-d-printed/

Bushey, R. (2014) "Researchers are making a 3D printer that can build a house in 24 hours," *Business Insider, Australia*, Jan. 10, http://www.businessinsider.com.au/3d-printer-builds-house-in-24-hours-2014-1

Clarke, B. (2011) "B.C. Camouflage Maker: The Invisible Man," *Globe and Mail*, October 19

Colvin, G. (2008) "Here It Is. Now, You Design It!," *Fortune*, May 26, p. 34

Cooper, S. (2012) "High-Tech Retail Will Revolutionize Shopping," *Vancouver Sun,* August, 14, p. A9

DaimierChrysler (2006) "A Realistic Look at the Factory of Tomorrow," *Highlight Report 2*, pp. 16 - 18

Dezeen (2013) "Road-Ready 3D-Printed Car on the Way," March 7, http://www.dezeen.com/2013/03/07/road-ready-3d-printed-car-on-the-way/

Del Ciancio, M. (2006) "Virtual Manufacturing,"
Manufacturing Automation, March/April

Dodson (2013) "New Microbatteries Combine the Advantages
of Lithium-Ion Batteries and Supercapacitors," *Gizmag*, May
8, http://www.gizmag.com/3d-microbattery/27162/

Dyer, N. (2008) "Bike Like an Egyptian," *Popular Science*, January

Eisler, T., Todd, B., Hanes, F. (2007) "Synchronized Manufacturing,"
Manufacturing Automation, September, p.12-13

Farsari, M., Chichkov, B.N. (2009) "Two-Photon Fabrication,"
Nature Photonics, Vol. 3, August, p. 450

Fortune (2009) "Under Armour Reboots," *Fortune
Magazine*, February 2, pp. 29 - 33

Globe & Mail (2007) "Pentagon Pays Dearly for
'Priority' Parts Shipping," August 17.

Globe & Mail (2011) "Japanese Parts Shortage Threatens
Global Auto Production," March 25.

Globe & Mail (2011) "Parts Shortage From Japan Idles Ford, Nissan Plants," April 2.

Globe & Mail (2012) "The Caterpillar Shutdown's Stark
Warning for the Industrial Heartland," Feb. 22, p. A8.

Gore, A. (2013) "*The Future – Six Drivers of Global Change*", Random
House, New York, NY, ISBN 978-0-8129-9294-6, pp. 30-33, pp. 241-244

Greb, E. (2010) "Thinking Inside the Box," *Pharmaceutical Technology*, May, p. 34

Greenberg, A. (2008) "Financing Fab," *Forbes Magazine*, September 1, p. 84

Jakovljevic, P. (2007) "Making Demand-Driven Manufacturing a Reality," *Advanced Manufacturing*, March/April.

Johnson, D. (2013) "3D Printing: Don't Believe the Hype," *Moneywatch*, June 21, http://www.cbsnews.com/8301-505143_162-57590222/3d-printing-dont-believe-the-hype/?tag=nl.e713&s_cid=e713&ttag=e713&ftag=

Krebs, B. (2011) "Gang Used 3D Printers for ATM Skimmers," krebsonsecurity.com, September

Lajoie, S. (1999) "300,000 Points of Light," *Forbes*, April 5, pp. 56-57

Lamb, E. (2012) "Cover Charge: New Spray-On Battery Could Convert Any Object into an Electricity Storage Device," *Scientific American*, June 28

La Puerta, W. (2014) "Chess Set.zip," Turbosquid.com, http://www.turbosquid.com/3d-models/3d-model-set-chess-pieces/408803

Lawn Rider, A. (2013) "IsoTruss Technology," *Inclusive Design & Mobility*, April 21, http://inclusivedesignmobility.wordpress.com/2013/04/21/isotruss-technology/

Lemley, B. (2000) "Behold, the 3D Fax," *Discover*, February, p. 26-28

Liedl, T., Hogberg, B., Tytell, J., Ingber, D.E., Shih, W.M. (2010) "Self-assembly of 3D prestressed tensegrity structures from DNA," *Nature Nanotechnology*; DOI: 10.1038/nnano.2010.107

MacGregor, A. (2012) "Leather Goods Firm Keeps Jobs on Shore," *National Post*, January 1

Mann, C.C. (2009) "Beyond Detroit: On the Road to Recovery, Let the Little Guys Drive," *Wired*, May 05

Manjoo, F. (2011) "Humanity's Obsolescence," *National Post*, October 11, p. A13

Maxey, K. (2013) http://www.engineering.com/3DPrinting/3DPrintingArtic
les/ArticleID/6661/Urbee-2-to-Cross-the-US-on-10-Gallons-of-Fuel.aspx

Media Lab (2013) Mediated Matter, http://matter.media.mit.edu/about

Military Wraps (2014) http://www.automopedia.
org/2009/08/17/10-amazingly-beautiful-stealth-vehicles-of-death/

Miller, C.C. (2012) "Celebrities Social Media Power,"
National Post, December 24, p. FP8

Mone, G. (2011) "Solar Alchemy – A 7 ft. High 3D Printer that Uses the Sun
to Transform Sand into Glass Objects," *Popular Science*, October, p. 73-74

Mone, G. (2013) "Microbial Powerhouses," *Discover*, April, p. 16

MPMN (2012) "CNC Machining of Small Parts, Laser Micromachining
Services," *Medical Product Manufacturing News*, May, p.16.

Nanoscribe (2014) "3D Photonics," *Nanoscribe*, http://
www.nanoscribe.de/?id=439&language=en

Nature (2014) "3D Printing Goes Nanoscale," *Nature* **507**, 277, March 20.

Okafor, E. (2011) "Business Models for Fab-Labs, *Timbuktu Chronicles*, April 5,
http://timbuktuchronicles.blogspot.ca/2011/04/business-models-for-fab-labs.html

Oxman, N., Kayser, M., Laucks, J., Gonzalez-Uribe, C.D., Duro-Royo,
J. (2013) "CNC Deposited Silk & Silkworm Construction," *MIT Media
Lab*, http://matter.media.mit.edu/environments/details/silk-pavillion

Pilieci, V. (2010) "3D Revolution Sweeps World of
Product Design," *Vancouver Sun*, December 20

Prahalad, C.K., Krishnan, M.S. (2008) "The New Age of Innovation – Driving Co-Created Value through Global Networks," *McGraw - Hill*

Ross, P.E. (2004) "Viral Nano Electronics," *Scientific American*, October, pp. 53-55

Shingler, B. (2014) "How a Micro-Magazine Smaller Than Salt Shows 3D Printers Big Potential," *Globe & Mail*, April 25.

Siscia (2008) "Fashion House Siscia Presents 3D Body Scanning Technology," http://www.siscia.com/hr/3dbodyscanning.html

Snieckus, D. (2010) "Window of Opportunity," *Renewable Energy News*, August 20, P. 21.

Suri, R. (2001) "Quick Response Manufacturing – Cut Lead Time and Roar Ahead of the Competition," *Advanced Manufacturing*, May, pp. 15-19

Teschler, L. (2013) "New Way to 3D-Print: Suspend Objects in Gel," *Machine Design*, August, http://machinedesign.com/3d-printing/new-way-3d-print-suspend-objects-gel?utm_source=feedburner&utm_medium=feed&utm_campaign=Feed%3A+pentoninteractive%2FMachine+%28Machine+Design+-+Articles+and+News%29&utm_content=Yahoo%21+Mail&Issue=MD-04_20130808_MD-04_41&NL=MD-04&sfvc4enews=42&YM_RID=tinarip%40yahoo.com&YM_MID=1415251

Tinari, P.D. (2000) "New Industrial Paradigm to Sweep the World," Office@Home Magazine, Spring, p. 7 - 8.

Wittbrodt, B.T., Glover, A.G., Laureto, J., Anzalone, G., Oppliger, C.D., Irwin, J.L., Pearce, J.M. (2013), "Life-cycle economic analysis of distributed manufacturing with open-source 3D printers, *Mechatronics*", http://dx.doi.org/10.1016/j.mechatronics.2013.06.002

Z Corp (2005) "Z Corporation 3D Printing Technologies," http://www.zcorp.com/documents/108_3D%20Printing%20White%20Paper%20FINAL.pdf

Chapter 3
The Second JOOM Revolution
Just-On-Order-Medicine

Introduction

If the 20th Century could be considered to be the era of mass medicine, the 21st will witness the full flowering of what will become known as *individualized* medicine. In mass medicine, the same drugs were dispensed to every patient who suffered from a particular ailment. For example, everyone with a mild headache was commonly given Aspirin, Ibuprofen or Acetaminophen, regardless of their different individual characteristics such race, sex or age. In the JOOM era, custom made drugs will be designed to treat a particular ailment while also taking into account a broad spectrum of characteristics particular to each individual. The transition to JOOM will have an impact on all aspects of medicine and this chapter will examine some of the most important changes that this will have on health care.

Synthetic Biology

This rapidly emerging area of research could be defined as the design and construction of new biological parts, devices and systems as well as the re-design of existing, natural biological systems for useful purposes. Progress in this sector was enabled by the rapid advances that occurred in the area of human genome DNA sequencing and synthesis technologies.

At the beginning of this century, a number of landmark studies reported on how various arrangements of similar segments of DNA resulted in completely different gene regulatory networks with completely different functions. These highly significant studies showed that:

> *"...the functions of biological systems are more than the sum of their genetic parts, and is determined by the interconnectedness among parts rather than their independent functions. Moreover, they demonstrated that these emergent properties could be predicted using computational and mathematical analysis....Essentially, DNA "parts" could be reused to endow cells with different and predictable biological functions."* (Power et al. 2012)

In 2003 the Registry of Standardized Biological Parts was set up to make all known information freely available, to make possible the reuse, adaptation, standardization and sharing of genetic material.

Some of the successful projects in the area of synthetic biology included the development by a team at the University of British Columbia of a fungus that produced monoterpene, found to be effective in combating mountain pine beetle infestations, the development, by a team from the University of Calgary, of organisms specially adapted to cleaning oil sand tailing ponds, while a team from Queen's University developed genetically modified worms that could be used for bioremediation projects.

In 2010, a self-sustaining and self-reproducing bacterial cell consisting of a synthetically manufactured genome was successfully created.

While molecular biologists were mostly limited to "working at the periphery" by carrying out rather simplistic modifications to existing genes and genomes, synthetic biologists will eventually gain a detailed understanding of the workings of even the most complex biological systems such as chromosomes, cells, tissues, organs and eventually, entire organisms. Eventually it will be possible to construct entire Just-On-Order-Genomes resulting in the creation of the first truly synthetic viruses, cells and organisms that can be built for virtually any desired custom application.

A completely different aspect of synthetic biology involved using naturally manufactured molecules such as DNA to create a new type of rewriteable data storage module. Beginning with a chromosome made up of DNA, the genetic material inside the cell, a synthetic biologist inserted a short string of laboratory-created DNA into the existing molecule. An enzyme was then used to cut a module from the DNA, and this was then flipped around, and then reattached into the same slot in the DNA double helix. In this position, the module was used to represent a binary "1." When it was desired to express a binary "0" once again, another set of enzymes was used to reverse the orientation of the module, returning it to its original position.

The applications of this technology were under intensive study but it could lead to computers thousands of times more capable than those of today, or to the development of artificially constructed immune cells that could be programmed to reproduce a fixed number of times before they inactivated themselves, so that they would not initiate a dangerous autoimmune response within the body. (Brown 2012)

It was highly desirable in synthetic biology to be able to rapidly and efficiently sequence the entire human genome. The first generation of DNA sequencing machines required more than 1,000 supercomputers running 24/7 for more than half a year to decode one human's genome and the procedure cost more than $5 million in computer time, manpower and chemicals. A company named Pacific Biosciences Inc., located in Menlo Park, California developed a DNA sequencing technique known as Single-Molecule-Real-Time Sequencing (SMRTS) that was tens of thousands of times faster than anything available before.

Faster sequencing techniques promised to deepen understanding of how normal cells, tissues and organs developed and to reveal what went wrong at the molecular level when a person developed diseases such ALS, Alzheimer's or diabetes. The eventual goal of this company's research efforts was to be able to deliver a complete and accurate analysis of the complete human genome in less than a quarter of an hour and for a cost of less than $1,000. (Lauerman 2008)

In 2011 researchers at the University of British Columbia developed a device capable of conducting what was known in the field as "single cell analysis." All of the equipment that used to fill an entire laboratory was incorporated into a single computer chip. The tiny chip was capable of separating the cells, adding reagent to highlight certain specific genetic sequences, and evaluating the amount of fluorescent light given off by the reaction. There will be countless applications for this technology outside of synthetic biology including research into stem cells, cancer, aging and biomedical engineering. (Kladko 2011)

An important and controversial aspect of synthetic biology was the long sought capacity to create artificial life. As early as 2003, researchers synthesized the first artificial virus. Its DNA consisted of about 5,000 base pairs or nucleotides and tests showed that it behaved identically to "natural" viruses. A short while later, a polio virus consisting of 7,500 base pairs was created in the lab. These successes indicated that it would in principle be possible to synthesize truly dangerous infectious viruses such as Marburg or Ebola. It could also be possible to make a completely man-made virus that combined the infectious properties of several natural viruses, making it uniquely infectious and deadly. Most ethically troubling, it was discovered that it was possible to prevent an organism from being successfully vaccinated to protect against a particular virus. This meant that a truly unethical dictator could, at least in theory, unleash a viral infection on the world combining the most terrible effects of the viral 1918 flu and the bacterial Black Death of the middle ages and from which no one could be successfully vaccinated. The accidental way that this discovery was made is described briefly below.

An ideal form of birth control would be to use the immune system to stop fertilization. As early as 1997, researchers realized both the promise and potential controversy associated with using infectious transgenic viruses as "carriers" of immuno-contraceptive genes to immunize the host against its own reproductive proteins.

Researchers at the Australian National University conducted an experiment that employed a virus to introduce high levels of an egg protein into female mice, promoting an immune response that would lead to sterility. The hypothesis was that by infecting mice with the genetically altered virus that normally caused mouse pox, the mice would contract an attenuated case of the disease and present an immune response to the virus that would extend to the egg protein. This extended immune response would lead to antibodies attacking the mouse's own eggs, causing sterility. Hence, an effective form of pest control. When it became apparent that production of antibodies against the mouse egg cells was adequate in some mouse strains, but not in more resistant ones, it was decided to try to boost antibody production by using an engineered recombinant mouse pox virus. The scientists understood that there was a possibility of a lethal immunosuppressant virus, but they believed that the genetic resistance of the mice would be strong enough to withstand systemic effects.

Unexpectedly, over expression of the egg protein by the virus led to total suppression of mouse cell-mediated immune response to the mouse pox virus. All ten mice in the experimental sample died. Furthermore, when mice already vaccinated against mouse pox were inoculated with egg protein expressing mouse pox virus, they suffered a mortality rate of over sixty percent. This suggested that even immune memory responses instilled through vaccination were inhibited by the virus-coded protein and the result was a total suppression of the immune response. The virus subsequently replicated without control and the mice then died of mouse pox. Since mouse pox was a member of the same family of viruses as human smallpox, the results clearly demonstrated that genetic engineering had the potential to render all existing smallpox, as well as other vaccines, useless. In seeking a way to develop an effective birth control "vaccine" researchers had accidentally discovered that they had found a way to suppress the benefits of the immunity conveyed by vaccination. In other words, the researchers had discovered how to produce Just-On-Order-Mortality. (FAS 2001) This was yet another example of your JOOM destiny...

It was discovered that many parents, who themselves had been vaccinated as children and so had benefited from life-long protection from infectious diseases such as polio and measles, were choosing *not* to vaccinate their children. This was because of the mistaken belief that vaccines were in some way harming their children. Consequently, epidemics of previously controlled diseases were regularly occurring in schools across North America. It was proposed by some health officials that it would only be fair for such parents to have the benefits of their own vaccines "revoked" using the method described above, so that they could then suffer from the same preventable diseases that they were needlessly forcing their children to be susceptible to. JOOM justice....

In 2009, British medical researchers created the first artificial sperm cells from human stem cells. The original intention was to offer a treatment for male infertility that could be offered through in vitro fertilization (IVF) clinics. But the potential to produce artificial sperm immediately opened up a host of ethical questions. For example, the technique could allow men who were long dead to father children. Initially, it was not possible to create viable sperm from female stem cells, but eventually it should be possible for women to dispense with men entirely and to "clone" themselves by being fertilized with synthetic sperm made from their own stem cells, or from the stem cells taken from female partners and going on to incubate and to give birth to children who technically had no fathers. This situation could be termed Just-On-Order-Maternity. (Macrae 2009)

By 2012 Japanese researchers created synthetic sperm cells and combined them with artificially created egg cells to create a viable litter of baby mice. Both of these were made by starting with adult stem cells and transforming them into what were known as Induced Pluripotent Stem Cells (IPSCs). The procedure was cumbersome, with the egg stem cells having to be incubated in vivo inside the uterus of a living mouse until they transformed into viable eggs. When ready, these eggs were collected, fertilized in a test tube by the artificially produced sperm cells and then implanted into another surrogate mouse which brought the embryos to term and gave birth to them. The researchers had managed to create the first Just-On-Order-Life-Forms. (Chant 2012)

One of the potentials of synthetic biology would be to design new genetic translational codes and incorporate them into an organism's genome, thereby making the organism resistant to all viruses. Viruses enter cells and hijack their genetic machinery to force them to manufacture new viruses. This was only possible because both the cells and the viruses shared the same method of encoding information within the genome. But altering the genetic code of the cell in certain critical ways could, in theory, thwart the ability of a broad spectrum viruses from replicating. It could also, in principle, be possible to constantly repair the damage that an individual sustained to their DNA as a result of the stresses of daily life, thereby eliminating much of the degradation that eventually led to systems mutation, resulting in illness and eventually, death.

The next step would be the modification of the genome to facilitate, for example, the interfacing of a human being with mechanical and/ or electronic devices, leading to what has sometimes been termed a Cybernetically Engineered Organism. This eventually would mean that humanity could in principle gain control over its own evolution, accelerating it and directing it in ways that nature could not have foreseen. (Freeman 1960)

The technologies associated with synthetic biology were advancing so rapidly into unknown legislative territory, that in May of 2010 the US House of Representatives held hearings to gain a deeper understanding of both the risks and benefits associated with synthetic biology and genomics. Testimony from leading industry experts established that DNA-based of life forms as complex as human beings could be artificially designed by computer, manufactured in a lab and then inserted into a living cell to yield an autonomous, self-replicating life-form completely under the control of the synthetic genome. Both the US National Institutes of Health (NIH) and the Department of Health and Human Services (HHS) published screening frameworks for the designers of synthetic genomes to prevent ethical or societal abuses that could lead to the development of agents dangerous to human health. Perhaps the greatest fear voiced by the legislators and by the Department of Homeland Security (DHS) was the use of these technologies for bio-terrorism. The question that remained to be resolved was if sufficient controls could be put into place to reduce the probability of such future threats to society.

Tissue Engineering

In 2008 the US federal government created the Armed Forces Institute for Regenerative Medicine (AFIRM) that consisted of a large network of leading medical centers and universities whose mandate was to develop new techniques to successfully treat the defacing and maiming wounds often suffered by soldiers in combat. As an example, for a soldier that lost an ear to an Improvised Explosive Device (IED), the first step was to make a 3D computer model of the remaining ear, then creating a 3D titanium scaffold modeled on this ear and finally seeding it with cartilage cells taken from the patient's nose or ribs. These cells were then allowed to grow for a few weeks in a Petri dish. When they had completely covered the scaffold, a skin graft was taken from the patient and used to completely cover the cartilage. The ear had to then be buried under the skin so that the cells could be sustained until blood vessels would grow into the tissues. The concluding step was to stitch the new living ear into its rightful position onto the side of the head. It was also found that it was possible to grow the extracellular matrix, the connective tissue that holds cells together inside a muscle, allowing patients to re-grow their muscle mass after it was destroyed in battle. (Marchione 2012, Catalyst 2012)

Stem cells were immortal progenitor cells that divided endlessly, generating new tissues throughout an individual's life. They could be found in a number of locations within the adult body including lying dormant in the bone marrow, blood, adipose tissue and in the retinal pigment epithelium, behind the retina of the eye. Another more controversial source of stem cells was from human embryos. First isolated by researchers at the University of Wisconsin in 1998, these undifferentiated pluripotent cells gave rise to all of the tissues and organs in the body.

A breakthrough occurred in 2007 when it was shown that adult skin cells could be made to revert to their embryonic state. These reverted or Induced Pluripotent Stem Cells (IPSCs) were capable of transforming into a broad spectrum of different tissues. In 2008, a method was developed that allowed for the generation of an entire stem cell line from a single embryonic cell, eliminating the continuous necessity of killing embryos to extract stem cells. Stem cell treatments were used in experimental treatments for leukemia, sickle-cell anemia, Parkinson's, Alzheimer's, paralysis, blindness, scleroderma, HIV and many other diseases.

In 2012 scientists from a number of nations formed a collaborative venture to research the properties of stem cells. It was discovered that there was a master control gene that served to "turn on" blood stem cells, opening the possibility of generating a virtually unlimited supply of fresh blood. This revealed a whole new treatment option for individuals suffering from blood-based disorders such as leukemia. (Taylor 2012) It was also found that changing only three or four genes could cause skin cells to revert to stem cells, the state that they last experienced when the organism was an embryo. Previously, the technique for doing this was to use a virus to carry the modified genes into the target cell, but this sometimes had the unfortunate side-effect of inducing tumours. The new approach was to use RNA, the molecule used by the cell to carry out DNA's instructions. It was observed that RNA from the four critical genes had the effect of transforming target skin cells into intermediates that could be transformed into other tissues such as muscle, nerve or heart cells. This work brought researchers very close to the point of mastering the ability to generate Just-On-Order-Tissues at will.

As already mentioned, the work carried out by the AFIRM showed particular promise. A therapeutic technique was developed consisting of an extra-cellular matrix that activated the stem cells that were present in the adult body. The powder provided a scaffold that attracted and activated a patient's own stem cells to cause the re-growth of damaged, or even lost body parts. But the re-growth was always partial. To accomplish the complete restoration of a lost limb, it was necessary to recreate the complex molecular signalling mechanism that was present in embryos that was critical to regulating the growth of the stem cells. It should be remembered that cancer cells were essentially cells whose signalling mechanism had failed so that they began multiplying endlessly without control. (Lenzer 2009)

A California company named Vet-Stem was set up in 2002 to use stem-cell therapy to treat horses with various soft-tissue injuries. Unconstrained by the many stringent regulations that restricted the administration of new medical treatments to humans, veterinarians were freer to experiment with the use of stem-cell treatments for a wide variety of soft-tissue injuries. The results of such animal experimentation had direct applications to humans since the basic tissue structure was somewhat similar. The issue was of vital medical interest since more than 80,000 people tore their ACLs (an important stabilizing ligament in the knee) every year. The technique used mesenchymal stem cells extracted from the connective tissues and from the bone marrow. These were cultivated in a growth medium and then concentrated in solution. When injected into the body, they were able to regenerate injured muscles, tendons, ligaments and cartilage. The technique proved so successful at healing injuries in race horses, that medical doctors were encouraged to begin trials to test its effectiveness in humans. (Anthes 2011)

The first experiments with tissue engineering used the simplest of human tissues, namely, skin. At MIT in the late 1970s a researcher called Eugene Bell extracted cells from skin that could reproduce in a collagen gel, forming a living *milieu*. By the early 1980s he noted that cells extracted from the skin's epidermal layer, placed in the proper environmental conditions, could be made to grow, forming a tissue that was roughly equivalent to skin. A company named Organogenesis was established in 1985 to commercialize this new technology. The company's first commercial product was called Apligraf and was approved by the FDA for use in humans in 1998.

The company encountered some controversy when it announced publically that its main source of fibroblasts (cells that form the dermis) and keratinocytes (cells of the epidermal layer) was from foreskins taken from infant males who had not consented either to the surgery to remove their perfectly functional foreskins, or to donate this valuable tissue to a commercial entity that was going to generate huge profits from the medical products made from the tissue. These foreskin cells were especially desirable, because of their awesome superior reproductive capacity. (Parenteau 1999)

In the late 1990s the company Advanced Tissue Sciences created two commercial synthetic skin products called TransCyte and Dermagraft. The latter product was the first human tissue engineered product to receive regulatory approval by the U.S. Federal Drug Administration (FDA) for treatment of third degree burns in humans. This product was also extracted from human infant foreskins that had been taken without consent.

A company named Tengion located in Philadelphia pioneered techniques for the creation of implantable organs. The process was to take some cells from the target patient and to grow them in a culture for about a week on a biodegradable scaffold. Since the cells were extracted from the actual patient getting the implant, there was no threat of tissue rejection. Once the organ was fully developed, it could then be implanted into the target's body, to begin its new job as a new human organ. Because of their relative simplicity, bladders were the first organs to be cultured in this way, but there appeared to be no reason why this technique could not be used for other organs as well. (Halley 2009)

By the early years of the 21st century, advances and substantial cost reductions in 3D printers made it much easier for researchers to use them in various organ printing experiments. With thousands of scientists around the world working in the field, one by one the barriers to creating viable human tissues using rapid prototyping machines began to fall. For example, in 2012 an organization called Tissue Mircofabrication Lab developed a cost effective method of building a functional 3D blood vessel network, an absolutely vital step for the creation of viable human tissues. The first generations of commercially available 3D Bio-printers were generally unable to recreate the complex network of blood vessels that were found in natural tissues. (Adafruit 2013)

The technique was first to create a 3D CAD model of the vessel network, and this was then used to print out a mould (or negative) of this network made of molten sugar. The sugar was allowed to harden and then the organ tissue cells, kept viable in a bio-gel, were poured over the 3D sugar network. The gel was allowed to harden and then the sugar was dissolved, leaving viable engineered tissue that was ready for implantation. It was found that this technique offered a very fast turn-around from the initial computer model to final living tissue. (Aslam 2012, Graziano 2014)

Early 3D bioprinters often damaged the living cells during the ejection phase, undermining the viability of the process. Researchers at the Institute of Biological Chemistry, Biophysics and Bioengineering at Heriot-Watt University, Edinburgh, developed a valve-based bio-printer that was able to control the volume of each droplet ejected from the nozzle so that pluripotency (viability) of the cells was not affected. (Faulker-Jones 2013) Instead of ejecting individual stem cells (which were not viable on their own) the printer produced embryonic stem cell spheroid aggregates that were tiny spheres consisting of hundreds of stem cells that had a much greater probability of producing viable tissues. Another innovation was to have two feeds to the printer nozzle – one with cells and the other without, allowing cell aggregates of different sizes to be produced in the final printing fluid.

There were a number of people who questioned the ethics of using embryonic stem cells that required the killing of early-stage human embryos. One alternative was to use iPS cells that were taken from adult cells such as skin cells, then regressing them back to the embryonic stage, making them identical to embryonic stem cells.

The issue that still had to be addressed was how to develop an efficient means of producing 3D stem cell tissue and then to easily get it to self-convert into the target tissue, such as liver, kidney or heart tissue.

Liver tissue was especially desirable, because it was ideal for testing new drugs and for disease modelling. The ultimate goal remained to move beyond the printing of tissues to the production of entire organs, but this would require the 3D printer to be capable of ejecting constantly changing cell types as the print went from one 3D location of the organ to another, while at the same time building into the tissues delicate 3D structures that would eventually serve to supply blood to the tissues. Ultimately, the desirable final product was an exact replica of the patient's original organ that was ready for implantation. Another goal was to miniaturize the printer head so that it could be placed inside the body and used to directly repair damaged tissues anywhere that they had occurred in the body.

Unlike skin or bone, heart tissue could not re-grow after it was damaged by a constriction in its normal blood flow. Instead, there always remained a scar called an infarct that would not undergo normal synchronous contractions like healthy myocardial cells. Consequently, by the 1980s, with thousands of heart attack victims waiting fruitlessly for heart transplants, there was an urgent medical imperative to find ways of growing new heart tissue that could be used to patch over and repair infarcts. Compared to other bodily tissues, heart material was incredibly complex, incorporating physical and neural connections that conducted the electrical signals that allowed the fibres to contract coherently and vasculature that could efficiently deliver oxygen to all of the cells in the tissues. (Cohen et al. 2004)

There were a number of approaches used to fix damaged heart tissues. The simplest one was direct cell injection of stem or precursor cells. While this offered an easy method of delivery, the cell survival rate was low and so it was difficult to produce viable tissues. A second approach was to culture tissues in thin sheets that could be used to patch the damaged areas of the heart. While the sheets were relatively easy to grow in a glass dish, the tissues lacked vascularization, so they tended to be thin and fragile. A third approach to engineering new heart tissues was to first build a scaffold made from specially solidified alginate, a polysaccharide obtained from algae. This scaffold was then seeded with specialized stem cells that eventually grew into living tissues that were rich with vascularization. The scaffold eventually broke down and dissolved away, leaving the self-supporting living tissue.

As good as this approach was, the field will most likely be dominated by the fourth approach to tissue engineering, the one using 3D-tissue printing. The tissue cells were first suspended in a hydro gel and nutrient solution. This solution was then dispensed layer by layer by the nozzle of the 3D-printer onto a biodegradable scaffold that had been pre-moulded into the shape of the organ to be created. The big advantage with this technique was that tissues could be printed with all of the necessary vascularization already in place. This approach will eventually be developed to the point where any organ could be created from scratch. After printing was completed, the living cells eventually grew together to form independently viable tissues and the now functioning organ could then be implanted into the body by any competent surgeon.

An imaginative approach to making new hearts was to use animal ones as the preliminary scaffolds. The technique was first to chemically wash away the cells leaving only the protein/carbohydrate extracellular matrix. The next step was to incorporate cardiac progenitor cells harvested from the human patient and then to incubate the whole structure in a bioreactor. The result was a beating heart that would not suffer rejection because it was made from the patient's own cells. It should also be possible using this technique to make living "patches" of cardiac tissue that could be sewn over damaged areas of the heart to allow it to contract more strongly. The weak link in the process remained the extraction of the progenitor cells and so this remained an area of intense research. (Schaffer 2008)

The technique of using living donor tissue as a primary scaffold to support stem cells as they grew was gradually extended to growing many other tissues. In one case, a section of vein was taken from a deceased donor and all of the cells chemically stripped away, leaving behind just a tube shaped scaffold. Stem cells were taken from the patient's bone marrow and were used to grow enough tissue to completely cover the scaffold. After about two weeks, the now completely functional, living blood vessel was transplanted into the patient to replace a vein that had developed a severe blockage. A similar approach was used to make a wide variety of new tissues and tubular organs such as new trachea (windpipes), bladders and urethras. The great advantage with this particular technique was that since the new tissue was created from the patient's own cells, it was not necessary for them to take immunosuppressive drugs for the rest of their lives. (Cheng 2013)

Hundreds of millions of people around the world suffer from the chronic effects of Type 1 Diabetes. Researchers found that it was possible to direct stem cells to mature into insulin-producing pancreatic beta cells and then to inject them into the body. Once in the body these cells could, on their own, sense glucose levels in the blood and secrete the appropriate amount of insulin. When tested in diabetic mice, this technique essentially "cured" them of the disease. The hope was that with additional research to perfect the technique that it could be made to work effectively in diabetic humans.

The design and creation of synthetic organs and devices that were designed to expand human abilities was an area of science known as Cybernetics. According to an article that appeared in Nano Letters:

> "*This field has the potential to generate customized replacement parts for the human body, or even create organs containing capabilities beyond what human biology ordinarily provides.*" (Nano Lett. 2013)

The introduction of 3D making into tissue engineering allowed for the first time the creation of hybrid tissues composed of living cells interlaced with printed electronics so small, that it formed a delicate network that passed between the cells. Such wired tissues equipped with this printed nanoelectronic scaffold could include sensors designed to detect irregularities within the body and then to send a signal to another device that could fix the problem before it became serious. This work marked the beginning of the creation of Just-On-Order-Cybernetically-Engineered-Organisms, work that held the promise of altering the course of human evolution.

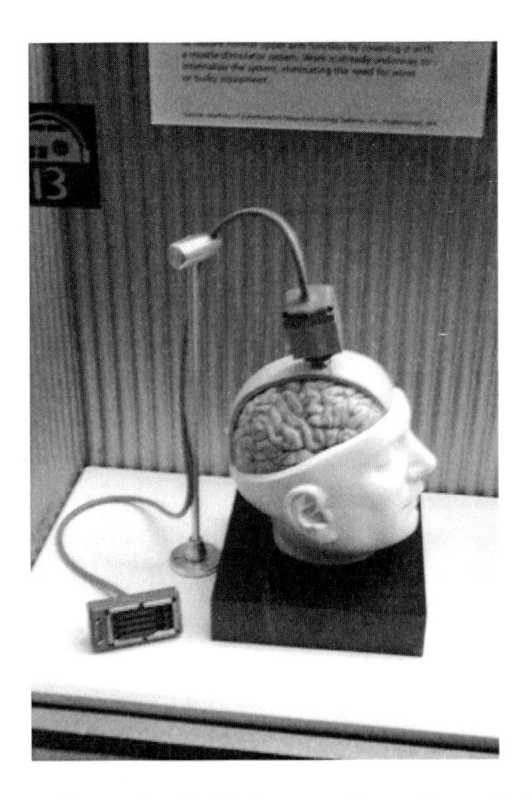

Figure 3-1: Example of a first generation cybernetic brain/
computer interface. In the future, such interfaces will be
wireless and miniaturized to the point of invisibility, driving
humanity closer to the predicted Singularity Point, the
ultimate merging of man & machine. Source: PaulWicks

In May of 2013 it was reported that researchers at Princeton University had used an inexpensive 3D printer to intermesh living cells with the electronics necessary to produce a superior response to sound. After the ear was printed out, the cells were cultured for several weeks so that they could grow together to form integral cartilage tissue that included the incorporated electronics. If implanted into a human, the printed bionic ear could hypothetically allow the recipient to hear frequencies more than one million times higher than normal biological hearing. (Sullivan 2013) This research elegantly linked together the formally separate fields of tissue and electrical engineering into a completely new area of hybrid research that ushered in a new era of super-smart implants, synthetic organs and high performance prostheses.

At Cornell University's Creative Machines Lab (CML), researchers produced a new open source 3D printer known as Fab@Home that could be assembled by anyone using the designs helpfully posted on-line. By combining living cells with a specially developed nutrient solution, it was possible for this machine to produce viable human tissues such as skin, cartilage (useful for fixing damaged knees) and even viable spinal disks. It was also possible to produce the porous biodegradable "scaffolding" that was necessary to construct complex organs such as hearts and livers.

3D bio-printing was especially successful at replacing damaged or missing bone. A patient who had more than seventy-five percent of the bone of his scull damaged in an accident, had all of it replaced by an exact duplicate of the damaged bone sections produced by a 3D bio-printer. A woman who had most of her lower jaw damaged by an infection had a replacement mandible made by a 3D printer using sintered titanium as the raw material for the core of the implant, and that was coated with a bio-ceramic artificial bone. Orthopaedic surgeons saw 3D making as being a significant paradigm shifting game changer in their industry. (Richmond 2012, Tan 2013)

Scientists in Britain succeeded in creating new brain tissue from human skin cells. Previously, embryonic stem cells had to be used to generate brain tissue, which by necessity, required killing the embryo. Researchers at the University of Cambridge found that it was possible to re-assign adult human skin cells so that they would grow into cerebral cortex brain cells. This development offered new hope for treatments for a host of neurological disorders including Parkinson's, Alzheimer's, stroke and epilepsy. (Gray 2012a)

Researchers at the University of Cambridge developed a machine based on Lego Mindstorms robotics that used hydroxyapatite-gelatin composites to automate the process of making artificial bone. Once programmed, the robot would do the same series of precise moves over and over until the complete bone structure was completed, allowing the construction of hundreds of replacement bones in a relatively short time and at low cost. The era of Just-On-Order-Bones had finally arrived.

Highly purified, high density alumina was commonly used for load bearing implants such as hip and dental replacements. Some of its desirable properties included bio-compatibility, durability and a low coefficient of friction. In principle, it was possible to produce alumina ceramic duplicates of virtually every bone in the body. Zirconia had greater toughness and lower deformability than alumina. It was ideal for creating delicate structures such as inner ear bones and bone anchors. Calcium phosphate-based bioceramics were commonly used for dental implants, jaw bones, and for facial and spinal surgery.

A company named Javelin located in Utah, invented a method for producing synthetic structures that closely mimicked bone bio-structures. The process started with the generation of CAT Scan (CT) data of the internal structures of the body. Software was then used to process the CT data and then to generate a detailed 3D computer model of all of the patient's internal anatomy through high-resolution surface triangulation, that could then be exported into the standard rapid prototyping, STL file format. The output started with sheets of green (unfired) bio-ceramic material. The software controlled a laser that cut the first cross-section from the sheet and the process was repeated, layer-by-layer, until the part was completed. The excess material surrounding the part was kept in place to help provide support as it was built. Before it was fired, the part had the general appearance of unfired parts created by traditional ceramic injection moulding technologies. The final step of the process was to remove the tape-cast binder and to sinter the part in a conventional furnace. The company used this process to build alumina, zirconia and phosphate-based mimetic bone bio-structures directly from patient-specific data. (Emory 2002)

Thousands of women underwent various treatments for breast cancer last year. The post-surgical appearance of the breast could range from a indentation the size of a cherry after a successful lumpectomy, to the complete removal of the entire structure after a radical mastectomy. There was an urgent unmet need for a restoration technology that could restore the original appearance of a cancer survivor's breasts.

A company named Cytori discovered that human fat contained stem cells that could be separated, cultured and then injected into the affected area. The injected tissue bonded with the surrounding healthy tissue, and new capillaries, lymph vessels and nerves gradually grew into the injected tissue. After a few weeks the new tissue was indistinguishable from the original breast structure. Research and clinical trials indicated that not only could this technology be used to restore the normal appearance of breasts after cancer surgery, but it could also be used for to help women who were unhappy with the size of their breasts and wanted to experience a more "natural" augmentation than was traditionally available with silicone implants. It was expected that this technique could be applied to repairing damage to other organs such as healing hearts damaged by infarctions and fixing lungs after malignant tumours were removed. There was even on-going research investigating the possibility of using fat extracted stem cells for penis enlargement applications or for the first time allowing circumcised men to regenerate their amputated foreskins.

People will soon be able to enjoy the body that they always dreamed of having instead of the one that they were born with, thereby becoming an integral part of your JOOM destiny...... (Begley 2010)

Information Technology (IT) Applied to Medicine

Few areas of modern technology have progressed as far or as fast as information-tech (IT). Revolutionary changes in IT dramatically changed the way information could be generated, analyzed, transmitted, stored, searched and retrieved. IT was responsible for the gradual conversion of medicine from a mostly reactive to a preventative science by making available to health care professionals all of the detailed genetic information needed to custom tailor drugs and treatments to individual human beings.

It was possible to identify specific biomarkers on an individual's DNA that could indicate that person's predisposition to developing a particular disease. (Natalizio 2008) Since there were over 30,000 genes and literally billions of possible interactions between them, combined with the fact that each individual had a unique genetic composition that could make them a preferred mark for a variety of different diseases and disorders, the amount of information that had to be collected and analyzed was truly mind boggling. But rapid advances in both IT hardware and software managed to reduce this complex analysis to a relatively simple daily task. A complete knowledge of each individual's detailed genetic information promised to dramatically improve both diagnosis and treatment, providing a significant enhancement to patient health care outcomes.

The company BioDigital Systems developed a computer-based animation and 3D visualization system that was so detailed and accurate that it could be used to model functioning medical implants within the human body. The sources of data used to create these hyper-accurate models was taken from hundreds of CT and MRI scans. The result of this research was several thousand detailed models of the human body. These models were available through the web site: www.biodigitalhuman.com The service offered by BioDigital was to allow makers of medical devices to have a detailed virtual picture of the functioning implants and to show in detail how they were interacting with surrounding biological structures.

Ex-Situ Organs

Most people saw cancer cells as universally evil and worthy only of disposal in a medical waste incinerator after being excised from the body. It therefore came as a complete surprise to many to hear that in 2009, researchers at the University of Chicago Medical Center, desperate to find treatments for patients who developed acute liver failure, developed a unique artificial liver machine known as the Extracorporeal Liver Assist Device (ELAD). What was unusual about this machine was that used almost half a kilogram of living human cancer cells to remove toxins from the blood of a patient with liver failure, while at the same time synthesizing vital proteins that they needed to survive. What prevented such a machine from being developed earlier was the simple fact that normal human liver cells died very quickly after they were removed from the body. An insightful researcher then had a creative insight that led to a solution to the problem. He realized that cancer cells were essentially immortal and that they could easily be made to replicate outside the nurturing security of the body. What followed was essentially just a matter of clever engineering, to find a way to separate out the plasma (non-cellular) part the blood, pass it over a porous membrane covering a chamber isolating the cancer cells, allowing the toxins to diffuse inwards to the cells to be treated, while allowing proteins such as Factor 8 (FVIII - blood clotting agent) to diffuse into the plasma from the cells. The goal of this device was to take over for a patient's liver function until a suitable liver donor could be found or until normal liver function could recover. (Stone 2009)

Pharmaceutical companies would love to be able to test the effects of new drugs on various human organs by testing them on a living person, but that would be highly unethical. So they were forced to experiment on living animals, but there was a high probability that these would not respond to the drug in the same way that human beings would. Another alternative was to test the substance on small tissue samples that had been altered so that they will not die like normal cells. Unfortunately, it was found that such cells would not respond to drugs in the same ways as the cells would if they were actually part of a living organ system within the human body.

It eventually became possible to build a microchip system that could accurately mimic the biochemistry and other behaviours of whole living organs. Such "Organs-On-a-Chip" were designed to accurately imitate virtually any particular tissue type and therefore they held the promise of eventually eliminating all animal testing and to be able to determine the impacts of new drugs on humans without actually exposing anyone or anything to harm. (Moyer 2011)

Researchers successfully constructed systems that imitated the response of the human lung, liver, intestine and breast to various proposed cancer drugs. As the technology was perfected, it was hoped that it will be possible to mimic increasingly complex human biological systems, eventually cumulating with the ability to predict the response of the entire human body to a particular drug or stimulus. The ultimate goal will be to develop the ability to manufacture a drug custom designed for the genome of a specific individual and to be able to test that drug on a synthetic biological system that would exactly mimic the real-life response of that particular person to the treatment, before it was actually given to them. This would be a giant step forward on the path leading to the era of personalized, Just-On-Order-Medicine.

Just-On-Order-Organs

In 2012 it was reported that a nine-year-old girl underwent an incredible six organ transplant operation that included her stomach, liver, pancreas, esophagus, small intestine and spleen. The organs had been taken from a recently deceased child who was of the same blood type and size. (Vancouver Sun 2011)

While this particular young girl fortunately made a full recovery, the unfortunate reality was that in the era of mass medicine, people were literally dying while waiting for suitable matching organs. Early in June 2013, the National Post reported that a US federal judge allowed one particular dying child to move to the front of the long organ transplant waiting list, creating a serious controversy. (Lauran 2013) A typical reaction to this decision came from Dr. John Roberts, the Chairman of the Organ Procurement & Transportation Network board, who commented: "...we can't build a system around making exceptions for everybody that isn't getting the transplant when they need it." Disparities in the availability of organs were a function of many factors such as the type of organ, but the greatest difference arose because of location, simply because in some places people donated more organs than in other geographic locations. What was to be done to bring desperately needed organs to those human beings who desperately needed them simply to survive?

A patient with kidney failure had to undergo regular daily dialysis treatments just to stay alive, hoping that a donor transplant organ might someday, maybe, perhaps, become available....please. Patients with bladder cancer would, after surgical removal of the diseased organ, have to spend the rest of their lives with urine collection bags attached to their belts. Even if a patient was successful in finding a compatible organ and if the transplant surgery was successful, they still faced a lifetime of a daily regime of anti-rejection drugs to keep their body's immune system of attacking the foreign organ and killing it. Clearly there had to be a better way.

The opening years of the 21st Century were an exciting time for experimental organ engineering. The original process for constructing synthetic tissues involved dipping a bio-degradable polymer scaffold into a bath containing a mixture of nutrients and viable cells. The technique worked reasonably well for producing small volumes of tissue or hollow organs such as bladders, but it was extremely time consuming and could not be used to make more complex organs such as livers, kidneys or hearts.

In 2010 a technique was perfected to allow the construction of viable organs such as human lungs. The method was to remove a lung from a donor and then to chemically strip all of the cells from the organ, leaving behind only its collagen-based scaffold. This inert structure was then placed in a nutrient-rich solution that would support cell growth. The solution was then saturated with adult stem cells taken from the patient. Before insertion into the patient, the stem cells had been programmed to become lung cells. These cells attached themselves to the scaffold and started to multiply, gradually forming themselves into the complex structures necessary to create a healthy lung. Stem cells programmed to become arterial and venial cells were then injected into the scaffold to form new veins and arteries. The final result was a completely functional lung that would never cause rejection problems for the recipient. (Zapana 2010)

Researchers at Stanford University discovered that a non-surgical technique that allowed a computer to construct 3D models of internal organs called Contrast-Enhanced-Electron-Beam-CT (CEEBCT) was able to spot heart defects with the same accuracy as invasive procedures such as angiography. The technique could also be used to construct accurate 3D models of a patent's organs, allowing 3D printers to construct exact copies of the organs using appropriately activated stem cells.

The ideal medical outcome would be to gain the capability to grow entire Just-On-Order, "ready-to-implant" replacement organs from scratch. What was required was a simple and reliable way to direct the growth of stem cells to become the desired target organ. Researchers found that with the addition of a single gene known as SOX17 into the DNA of one such stem cell, it was possible to direct the growth so that the result would be endoderm cells, the foundation of lung, liver and pancreas tissues. It was found that the addition of the master gene mimicked what happened in a developing embryo where the appropriate genes were successively turned on in a particular sequence. From the initial preliminary discoveries, the number of applications of stem cells proliferated widely. For example, researchers at the University of Sheffield were able to restore the ability to hear in deaf rodents by use of embryonic stem cells that had been modified to become otic-progenitor cells that were able to repair the damaged auditory nerve cells. (Ubelacker 2012)

While extreme levels of controversy were initially generated by the fact that stem cells were taken from human embryos, later it was discovered that the addition of four critical genes into ordinary human cells succeeded in converting them into cells that were virtually identical to human embryonic cells, but free of the ethical baggage. Most importantly, because it was a patient's own cells that were reprogrammed to become stem cells, the resulting organs would be recognized by the immune system as "friendly" tissue and so the patient would not have to take anti-rejection drugs for the rest of their life. (Abraham 2008)

At the University of Pittsburgh, researchers discovered that it was possible to transplant hepatocytes (liver cells) and have them survive in unusual parts of the body, such as in the lymph nodes. The result of this was the production of what could be called a distributed-network virtual liver. Experiments showed that when enough of the nodes were populated with transplanted liver cells, the cumulative total was to replace the functions of an entire liver. The lymph nodes were ideal places to grow and sustain liver cells, since once there, the cells had access to the blood supply, nutrients, and hormones as well as to the various signaling agents circulating through the body.

The technique began with adult stem cells that had been reprogrammed to mimic embryonic stem cells. These cells, known as induced pluripotent stem cells, (iPSCs) were then directed to become liver cells, and these would then be injected into various lymph nodes distributed throughout the body. The lymph nodes were considered to be bioreactors capable of harbouring cells from any organ that either produced new cells or that secreted hormones. One limitation with this technique was that there were still doubts that the transplanted cells could carry out all of the multitudes of functions of a healthy organ as complex as a liver, and there were legitimate concerns that the transplanted cells could sometimes migrate to other locations in the body such as in the brain or the heart and cause severe problems in these new growing sites. (Piore 2012)

Traditionally, if a patient suffered from an obstruction, it was necessary for the surgeon to construct a bypass using veins harvested from some other part of the body. A company located in San Diego named Organovo, working in collaboration with researchers at the University of Wisconsin at Madison, developed a 3D printer that was capable of producing lengths of synthetic blood vessels made from human cells without the need for any supporting scaffolding. The machine could precisely position endothelial cells to create the vessel walls, smooth muscle cells to mimic vessel expansion and contraction and fibroblast cells that created the outer structure found in natural vessels. This development meant that it was only a matter of time before unlimited lengths of vessels could be produced on order for any patient requiring it. (Calamia 2011)

The problem was that the more complex organs consisted of extremely intricate networks of blood vessels that fed nutrients to the internal cells. When attempting to grow such an organ, the internal vessels simply could not be constructed fast enough to keep the interior cells alive. This proved to an insurmountable problem until Thomas Boland of Clemson University and Vladimir Mironov of the University of South Carolina thought of using an ordinary ink-jet printer which had been re-adapted to spray a nutrient mix infused with living cells instead of ink. The printer was also modified to print in an additional dimension beyond the two-dimensions of a usual off-the-shelf machine. The cells were deposited by the printer nozzle into a biodegradable gel that held the cells in position until they could grow into each other to form a continuous living tissue. Using this printing approach to manufacture tissue allowed the network of blood vessels to be produced at the same time so that the cells could survive until they could grow into viable tissue. (Stroh 2003)

In 2003 researcher Makoto Nakamura of Toyama University in Japan developed a 3D printer that ejected human cells through a capillary tube less than a tenth of a millimetre in diameter. This printer head had the capability of ejecting cells into a specific position in 3D space with an accuracy of about one-thousandth of a millimeter and could print out a synthetic blood vessel at a rate exceeding one centimeter per minute. Nakamura first confirmed that the cells survived after being ejected from the printer nozzle. By extensive experimentation he then discovered that he could prevent the cells from drying out by placing them in an alginate sodium solution and that their integrity could be maintained after printing by ejecting them into a calcium-chloride solution.

In 2006 a medical researcher named Dr. Anthony Atala at the Wake Forest Institute for Regenerative Medicine (WFIRM) in North Carolina, developed a procedure for making human bladders and urethras. The method was to use precursor cells extracted from a patient's own bladder that were laid down on a biodegradable scaffold made into the required three dimensional shape. The precursor cells could be cultured in the lab to form the muscular cells that made up the outside of the bladder as well as the more specialized cells that made up its inside lining. Dr. Atala built the first artificial bladders by painting the cultured cells onto the scaffold and then immersing the whole structure into a nurturing growth medium. Warmed to body temperature, the cultured cells went on to grow, mature, multiply and to finally interlink to form healthy bladder tissue. The final stage was to implant the new bladder into the happy patient, who would not face a lifetime of organ rejection problems because the cells in the new bladder were their own. (Tam 2013)

Painting the cultured cells onto the scaffold was slow and tedious, so Dr. Atala next experimented with the use of a standard inkjet printer adapted to eject cells immersed in growth medium instead of ink. The printer was also modified to print in three dimensions instead of just two. Using this approach, the cells were laid down layer by layer onto the scaffold to eventually create a complete bladder.

Ongoing research by Dr. Atala at WFIRM involved finding ways to heal fresh wounds such as cuts or burns in hours instead of weeks. A laser scanner would first be used to analyze the shape and depth of the wound. This information would then be passed to the computer controlling an organic 3D printer. Layers of appropriate skin and tissue cells would be deposited layer by layer into the wound until the normal structure is completely restored.

Organovo (introduced previously) developed a machine that, starting with stem cells extracted from adult bone marrow and fat, was capable of printing out tissues such as skin, muscle and as described before, segments of blood vessels. The initial stem cells were persuaded to differentiate into the appropriate type of tissue by the addition of the specific growth factors. The eventual goal was to be able to print out organs of increasing complexity such as bladders, kidneys, livers and hearts that could be implanted into the body. Organovo's machine worked very much like a common off the shelf 2D inkjet printer, except with the added ability to print using living cells deposited layer by layer in a third dimension. Since the cells were alive, they merged with the cells surrounding them to form continuous functioning tissues. For constructing tubular structures such as blood vessels a sugar-based hydrogel was used as a scaffold to control the thickness of the living cell layer. When the maturation of the tissue was complete, the scaffold was pulled from the inside and peeled off the outside of the tube to create a ready-to-graft blood vessel. The long term goal was to develop the capability of printing organs *in-situ*, (in the body) thereby eliminating the need for a transplant operation.

Closely related to the creation of Just-On-Order-Organs, was the science of creating Just-On-Order-Body-Parts. The technology of human (and animal) prosthetics was advancing so rapidly that already there was talk about the need to discuss the implications of creating people with artificial limbs that not only equalled the capabilities of ordinary humans, but that far exceed them. The extreme controversy that resulted when an athlete with two prosthetic legs wanted to be allowed to run against able bodied athletes in the Olympics only hinted at the magnitude of the ethical dilemmas that would have Solomon himself holding his head in confusion. Clearly, it is only a matter of time before so-called "disabled" athletes using the next generation of prosthetic technologies would start logging performances that were far better than able-bodied athletes. Eventually in the interest of fairness, it will be necessary to restrict the use of advanced prosthetics just as performance-enhancing drugs were restricted by the governing bodies of various sports. (Chivers 2013)

What is certain is that prosthetics will increasingly be manufactured using 3D making technologies. The simplicity of the process was compelling: Scan the healthy limb or body part (if possible) to create a digital model, make a mirror image of the model, and then manufacture a synthetic copy of the body part. In some cases, it may not be necessary to make a prosthetic that actually looks like the body part that it is replacing for it to perform like that part. An example can be seen in running blades that allow handicapped athletes to run as fast (or even faster) than the able-bodied, even if they do not look like them.

The Era of Personalized Just-On-Order-Medicine

The U.S. spent about $2.6 trillion on healthcare in 2010 and was expected to need about $4.5 trillion to sustain the system by 2020. Clearly, this growth in expenditures was not sustainable and urgent steps had to be taken to prevent a general system meltdown.

Fortunately, there were many rapidly advancing technologies that were converging to move medicine from the era of mass treatments to ones tailored for each individual. It had become clear that the tactical and strategic uses of smart technologies were fundamental to being able to deliver Personalized-Just-On-Order-Medicine to every patient.

The overall goal was to improve the quality of care and to reduce costs throughout the system. These technologies included genomics, micro-sensors, wireless networking, cloud computing and social networking. A new field was emerging known as digital health, or the use of a wide variety of digital technologies for prevention, diagnosis, monitoring and treatment of disease.

The first component of personalized medicine involved new generations of microscopic, implantable sensors that could continuously monitor important bodily functions and transmit the information in real time to the individual's healthcare provider. For example, sensors could measure the galvanic skin response due to sweating, motion using a three-axis accelerometer, temperature and heat flux to determine the heat loss from the body. High degrees of accuracy were obviously paramount with these sensors and the goal was to achieve a reproducible accuracy that exceeded ninety-five percent.

The second component was the technology to transmit data from the implanted sensors to the external world. Bluetooth Low Energy (BLE) was successfully used to network a wide variety of e-health devices and systems. It worked well because most sensors only sent limited volumes of data and could operate at low power levels. Another system was the Ultra-Low-Power Medical Implantable Communication Service (ULPMICS) made by Microsemi that was used to link an implantable wireless radio frequency (RF) transmission chip to an external monitoring device.

The third component of personalized medicine dealt with technologies to receive and analyze the data from the sensors. An effective way of stimulating creativity in a particular domain was by sponsoring a competition that offered a nice big fat monetary prize. Both Nokia and Qualcomm sponsored competitions offering respectively $2.5 and $10 million prizes to teams that could develop external devices capable of non-invasively using all available sensing technologies to diagnose at least fifteen different benchmark diseases. In 2011 the USFDA approved the first smart-phone based blood glucose meter (BGM) developed by Telcare Inc. Units with ever increased capabilities began to be introduced henceforth to the market on a regular basis.

The fourth component of medicine designed to address the uniqueness of each individual was various dynamic medical implants such as adaptable cardiac pacemakers, drug and hormone dispensing pumps, autonomous defibrillators and deep brain stimulation devices. Eventually it would be possible to miniaturize virtually the entire instrument content of a typical medical clinic and implant it into the human body to both sense, diagnose and treat a broad spectrum of different illnesses and diseases. The eventual goal was to make the entire sensor, analysis and transmission system so small that it could be comfortably placed on the surface of, say, a contact lens, and it would be able to provide a 24/7 monitoring of a person's health by an analysis of their tears.

According to some studies, patients did not take from thirty to fifty percent of their prescribed medications, resulting in billions of increased systemic costs. One solution was to use internal sensors that would transmit signals to specially equipped smart phones running apps that would keep track of the time that (and if necessary, the place where) a drug was taken, the manufacturer of the drug, the dose of the drug in the body, heart rate, respiratory rate, blood pressure and many other responses, all as a function of time. The sensors could be powered by miniature batteries, blood sugar, stomach electrolytes, or by external electromagnetic coupling.

It has long been known that cancer patients with exactly the same diagnosis often reacted completely differently when given identical treatments. The different reactions were the result of each individual's unique genetic structure, so further advances in cancer treatments strongly depended on the ability to relate individual genomics to specific responses to certain treatments. Personalized medicine meant using an individual's unique genetic information to diagnose, prevent and treat illness. This would mean that the information taken from a single saliva sample, for instance, would be capable of indicating all of the health risks that an individual could likely face in their lifetime. As an example, for females who knew that they had a family history of breast and/or ovarian cancer, genetic testing could confirm if they were actually at a high risk of developing these diseases. What happened when women who were found to have mutations to their BRCA1 or 2 genes were refused health insurance?

The problem was that work has only just begun of systematically going through the more than 30,000 human genes and constructing a catalog of the variations that could occur in each one and then linking these variations to specific human health outcomes. Also the various genetic markers that have been linked to certain disorders do not provide an accurate estimate of the actual probability that a specific individual would actually suffer from the disorder. Even when all of the genes were identified that could be linked to a particular condition, that still left open the question regarding all of the non-genetic factors that also played a significant role in the condition. In many cases, incomplete information could be more harmful than no information at all. (Weeks 2012)

A critical component of personalized medicine was pharmacogenetics where an analysis of a person's genetics was used to determine a person's likely reaction to a proposed drug treatment. Among the numerous benefits of this new approach was the reduction of the unacceptably high rates of adverse drug reactions as well as avoiding treatments that were simply ineffective. Cells in malignant tumours were prone to DNA mutations and a detailed analysis of these changes could be used to custom tailor a particular drug treatment for each tumour. The genetics of liver cells could be used to determine how effectively a person could metabolize certain classes of drugs, allowing physicians to better calculate the appropriate dose of a specific drug treatment for each individual patient.

Until recently, it was not possible to keep tumour cells alive in the laboratory for very long. Then it was discovered that, by using techniques developed for stem cell research, it was possible to keep malignant cells alive for years outside the body. With the profound discovery that the characteristics of each person's cancer were unique to that individual, for the first time it became possible to test the effectiveness of chemotherapy drugs on tumour cells before giving them to the patient.

Like the other tissues in the body, blood was unique to each individual. Because of this, the chemical markers that were deposited in the blood after a heart attack were also unique. This meant that each person responded differently to the drugs given after a heart attack. The goal of synthetic biology was to make a virtual model of each patient's blood, so that each individual's reaction to a particular series of heart drugs could be predicted before the drugs were actually administered.

Triple Negative Breast Cancer (TNBC) was one of the most lethal forms of the disease and for decades it was treated as if it were only one unique type of malignancy. Then the genetic analysis of hundreds of such tumours indicated that in fact there were huge variations in the DNA of the component cells. The conclusion was that in order to be treated effectively, TNBC tumours had to be first, genetically sequenced, and second, treated by the most effective means known for the specific gene mutation in question. This meant that in order to progress, cancer treatment had to be personalized to a degree never before thought necessary, or possible. (Fayerman 2012a)

Each person's tumour was unique and it was important to gain a complete understanding of this uniqueness if treatment was to be successful. The problem was always that while breast structure was highly three dimensional (in most women), established imaging techniques produced 2D images that did not reveal significant details about certain tumours. This is why the invention of an imaging technology known as tomosynthesis was such a significant advance for personalized medicine. The new technology created a 3D image by taking a series of pictures as the X-ray emitter was rotated in a complete orbit around the suspect area. The image series was then submitted to a computer program that generated a clear 3D image of the tumour that revealed many more intimate details that enabled a professional to select the most appropriate treatment.

It was believed that many breast cancers started with a transformation of the cells found in the milk ducts, and such tumours were known as Ductal Carcinoma in Situ (DCIS). When small, such malignancies remained within the milk duct, and at this stage they were easily treated. However, as the tumour grew, it gradually invaded other parts of the breast forming increasing numbers of secondary tumours. These tumours secreted hormones that promoted the growth of blood vessels that assured a steady nutrient and oxygen supply to the malignant cells. The name given to this process was angiogenesis. It was known that not all DCIS progressed to the invasive stage, but the problem was how it could be possible to distinguish between the highly invasive DCIS from the one that would harmlessly remain in the milk ducts.

To protect patients from overtreatment, it was necessary to develop new Just-On-Order-Medical-Imaging technologies capable of differentiating between the different types of DCIS. One approach was known as Microbubble Ultrasound (MBUS), a procedure that involved injecting a tiny volume of bubble solution into a vein, and then using the bubbles to create a 3D profile of the blood vessel growth around the DCIS. A second approach was known as Ultrasound Tissue Elastography (USTE) that took advantage of the fact that the presence of malignancies tended to reduce the natural elasticity of tissues. By seeking areas of reduced tissue elasticity, USTE offered the opportunity to detect cancers much earlier, even in patients with dense breasts, which previously made accurate detection much more challenging. (Globe & Mail 2012)

Medulloblastomas (brain tumours) were a leading cause of death in children. Research into the genetics of childhood brain cancers revealed that in the majority of cases the tumours were only low-risk and consequently they were being over treated. It could be concluded that being able to better match the treatment with the tumour would not only benefit the patient but would also save the medical system millions of dollars annually in inappropriate treatments. The difficulty would be that parents would have to become receptive to a novel idea – the thought that less but better targeted treatment for their child was the way to go. (Fayerman 2012b)

The research involved examining the genetics of biopsies taken from thousands of tumours from all over the world. Certain biological markers were associated with the degree of pathology of certain tumours. A data-base of such markers could help ontology clinics deliver better, more accurately targeted treatments. The risks associated with un-targeted medicine were many including physical and cognitive impairment as well as an increased risk of metastasis (spreading of the original tumour). Being able to refer to the data base and to be able to exactly tailor the aggressiveness of the treatment would be a massive step forward in the battle against childhood cancers.

A number of cancer sufferers noticed that their companion animals often took a particular interest in the area of the malignancy, showing a fascination with the apparent odour of the tissues there. In fact the ancients recognized that the odour of a person's breath often could be used as an indication of disease. Hindu physicians noticed more than 3,000 years ago that the breath of diabetic patients often smelled sweet. The types of Volatile Organic Compounds (VOCs) given off by the body could be used as biomarkers of various diseases such as cancer. It was long known that all cancer was associated with a dysfunction in the energy metabolism of cells and that each type of malignancy gave off a different set of VOCs that could be identified. This meant that in principle, it should be possible to construct a Just-On-Order-Medical-Test of a patient's breath to determine exactly what type of cancer that they were suffering from. One technology being harnessed to accomplish this was known as Laser Infrared Sample Analysis (LISA) that made use of the fact that each molecule absorbed a slightly different wavelength of infrared (IR) light that allowed it to be identified. It was hoped that eventually more than fifty diseases would be identifiable just from an analysis of a patient's breath. (Pilger 2012)

It was well known that malignant cells were not always attacked by the body's immune system because they were not seen as "foreign" since they consisted of once normal cells that had gone rogue. Hence a very promising area of cancer research was in the development of vaccines that empowered the body's own immune system to attack and destroy only the malignant cells, while ignoring the healthy cells. The process could be termed Just-On-Order-Mucin 1 recognition, because it was discovered that the surfaces of cancer cells (but not normal cells) were covered with MUC1 molecules. The vaccine primed the immune system to recognize this molecule so that the cancer cells could be destroyed. (Gray 2012b)

The approach taken by researchers at the University of Pennsylvania to fight leukemia, known as Adoptive T-Cell Transfer, was to reengineer each patient's T-cells to specifically target a molecule known to populate the surface of malignant cells, but not normal ones. This solution could also be used to allow the body's immune system to target other cancers including those of the lung, skin and ovaries. Many questions still remained to be answered about this therapy, the most important of which was if the activated T-Cells would be persistent enough to continue to protect the body from disease should the malignancy return. (Beasley 2012)

The genetic mutations associated with a host of inherited diseases have been identified including Tay-Sachs, cystic fibrosis, Parkinson's, MS, diabetes, cancer, cardiovascular and psychiatric disorders. This was possible even though these diseases were found to be controlled by a complex series of genetic markers. Combined with this capacity, the Heart Institute at the University of Ottawa, working in collaboration with the biotech company Spartan Bioscience, developed a DNA test that could produce detailed results in minutes instead of the days that previous tests required. (Spears 2012)

A central tenant of Zen was the interconnectiveness of all things. Researchers at the Michael Smith Genome Sciences Centre at the British Columbia Cancer Agency discovered that the mutation of a single gene could be connected with the development of a number of cancers that previously were thought to be unrelated. The significance of that particular discovery was that it permitted the investigation into the molecular origins of cancer, rather than just focusing on the resulting tumours. Previously, it was believed that each class of tumour was caused by a different genetic mutation, but the recent discoveries showed that this was not the case. The gene in question instructed cells when to eliminate certain other genes. Apparently, when this ability was disabled, even partially, then the inability to eliminate the other genes led to the development of malignancy.

This research provided strong support for a systems view of biology, where focus was placed on the fundamental molecular composition of tumours, rather than on where they were occurring in the body. This research was made possible by the fact that for the first time oncology researchers had a complete genomic sequencing centre at their disposal, permitting them to conduct detailed investigations into the molecular basis of disease. When molecular level 3D printers become available, then it will only be a question of when and not if, it would be possible to reconstruct the damaged portions of an individual's genomic code, atom by atom, removing the faulty programming that would have led to them developing cancer. (Carman 2011)

Using what was known as a "Zinc-Finger" consisting of two protein strands bound together by a Zn atom, researchers demonstrated that it was possible to excise a gene that enabled the HIV virus to attach to T-cells. The goal was to replace all of the body's T-cell population with the altered cells so that, having nowhere to attack, the virus would simply die off before it could do any harm. Previously, individuals infected with HIV faced a lifetime of anti-viral therapy costing up to a quarter million dollars per year. The new treatment offered the advantage that the altered cells maintained the same effectiveness as normal T-cells at attacking pathogens that entered the body. This treatment resulted in unprecedented improvements in subject's T-Cell counts. (Bohan 2011)

A small percentage of human beings benefited from a genetic mutation that effectively made them immune to HIV infection. This mutation was first observed in prostitutes in Nairobi, Kenya. Despite being exposed to HIV many times a day, they never went on to develop the disease. In 2006 a man who had been HIV positive for more than ten years developed acute myeloid leukemia. After traditional chemotherapy failed to halt the progress of the patient's blood cancer, his oncologist decided to try a bone marrow transplant using stem cells taken from a donor who had the genetic mutation that had been shown to provide protection from HIV infection. To the amazement of everyone involved, the patient's viral load fell to zero and testing indicated that he was still HIV-free more than three years later.

This man was essentially the first person on earth to be "cured" of HIV infection, in addition to having his leukemia treated as well. The problem was that the procedure was highly risky, with a more than thirty percent mortality rate, so most HIV positive patients were simply treated with life-long antiviral therapy, which, while not providing a cure, at least didn't expose patients to a high risk of death. The aim of subsequent research was to find Just-On-Order-Medical approaches that would dramatically reduce the risk of donor stem cell transplants, while still achieving the same HIV cure. (Sheridan 2010)

Dramatic technological advances in micro and nano-sensors permitted the development of new generations of implantable medical devices that could autonomously monitor a broad spectrum of conditions inside the human body. These Micro-Electro-Mechanical-Devices (MEMS) could be used to monitor for the first indications of cancer, and then could be programmed to dispense a precise dose of a chemotherapy drug exactly at the location of the pathology. This approach was far superior to dosing the entire body with toxic drugs and hoping that the concentrations would be high enough to kill the malignant cells. These advances had been driven by a number of supporting research efforts into biomaterials, nano-electronics, ortho-biologics, neurology and systems biology.

Particularly exciting were the advances in deep brain stimulation using biomedical implants as effective treatments for Parkinson's disease, essential tremor and other movement disorders. These devices consisted of programmable micro-processors combined with Lithium Ion (Li-Ion) battery powered stimulators attached to an electrode. The electrode passed through a hole in the skull and was implanted into a specific targeted region of the brain.

In the 1960s the medical practice of cardiology consisted of diagnosis followed by drug prescription and monitoring. Starting in the 1970s however, a whole host of new technologies became available to cardiologists including arterial stents, electronic pacemakers and implantable defibrillators. In the 21st century, neurologists were similarly empowered by many new technologies that helped them treat many of the formally difficult, or impossible to treat brain disorders. (Rockel 2011)

In the emerging era of personalized medicine, mass treatments for a particular disease were abandoned in favour of custom designed treatments carefully formulated for each individual patient. To accomplish this, physicians had to develop the capacity to closely observe the evolution of disease by studying its unique genetic markers. This allowed them to make a custom diagnosis and then to recommend a therapy unique for each individual. This procedure was generally known as

Targeted Therapeutics, where specific protein products of a person's unique genome were used to develop a unique treatment for each person suffering from a particular disease.

As an example, the drug trastuzumab developed by Genentech, was designed to target the protein encoded by the HER2 gene. How well the drug would succeed at treating a particular patient suffering from metastatic breast cancer was possible to predict by the amount of expression of the HER2/neu protein. At the same time it was possible to test each individual patient for Multi-Drug Resistance (MDR), where the cells of an individual's tumour gained the ability to move the molecules of the drugs used in common chemo therapy treatments outside the cell before they could cause any damage. Detecting a high concentration of trans-membrane permeable glycopotein would indicate if a particular program of chemotherapy was likely to be effective for a particular patient, and would minimize the risk of chemotoxicity. (Bioscience World 2012)

Other components of personalized medicine included:

- **Pharmacogenomics**: The determination of a particular individual's potential interaction with a specific drug treatment at the molecular level so the adverse interactions between drugs could be minimized and so that custom dosages could be assigned

- **Theranostics**: The development of tests that could identify the best course of treatment for each individual patient and then providing feedback on how well that treatment was working.

- **Genomic Mapping**: Identification of which genetic mutations could interfere with a proper diagnosis in a particular patient (or group of patients) and the evaluation of the probable risk that a specific drug treatment would entail for that patient.

- **Analysis Using Multiplexing Microarrays**: Since most diseases were known to be caused by multiple genetic mutations, this provided the ability to test many genetic markers at the same time, quickly, accurately, efficiently and without the need for skilled labour.

Melanoma was one of the deadliest forms of cancer, killing more than eighty-five percent of those unfortunate enough to be afflicted by this skin disorder. Fortunately, promising research uncovered a personalized Just-On-Order approach to treatment. The approach was to start with a highly porous biodegradable plastic disk that was saturated with proteins extracted from a mouse tumour and with an inflammatory protein known to attract immune cells called GMC-SF (granulocyte macrophage colony-stimulating factor). Once the disk was implanted under the skin of a human patient, the body's immune system began mounting an attack against the foreign cells. But in doing so, the immune cells also unavoidably came in contact with the unique chemical markers on the tumour proteins, thus learning to recognize them.

With time, the trained immune cells made their way to the lymph nodes, a critical component of the body's immune system. There they exposed indigenous immune cells to the same foreign proteins, training them to recognize and attack any tumour cells that they would encounter. Eventually, countless millions of cells were trained in this way, leading to the creation of a powerful immunological response. By 2010, this experimental treatment led to the construction of a disk-based system that could cure more than forty percent of patients originally diagnosed with metastatic melanoma. The testing continued at the Dana-Farber Cancer Institute to improve the odds still further.

An area of research that showed a great deal of promise was "biologics," classes of protein based drugs that were derived from living cells. These were particularly effective at helping relieve the suffering of patients with rheumatoid arthritis but were also used in cases of cancer, heart disease, skin and spinal column problems. The main difficulty was the fact that these treatments were significantly more expensive than those using conventional drugs.

A major genomic discovery was the fact that events that happened in a person's lifetime, such as stress, poverty, abuse or simply poor nutrition could actually impact their DNA through a process known as methylation. It was found that DNA methylation could affect which of the approximately 30,000 genes were turned on or off, and that expression of these genes was linked to aging and susceptibility to numerous diseases such as cancers. Once the links between particular patterns of methylation and the expression of specific genes in the body were determined, it will be possible to create a Just-On-Order-Methylation process based on molecular 3D printers to specifically custom demand the expression of some genes while silencing others.

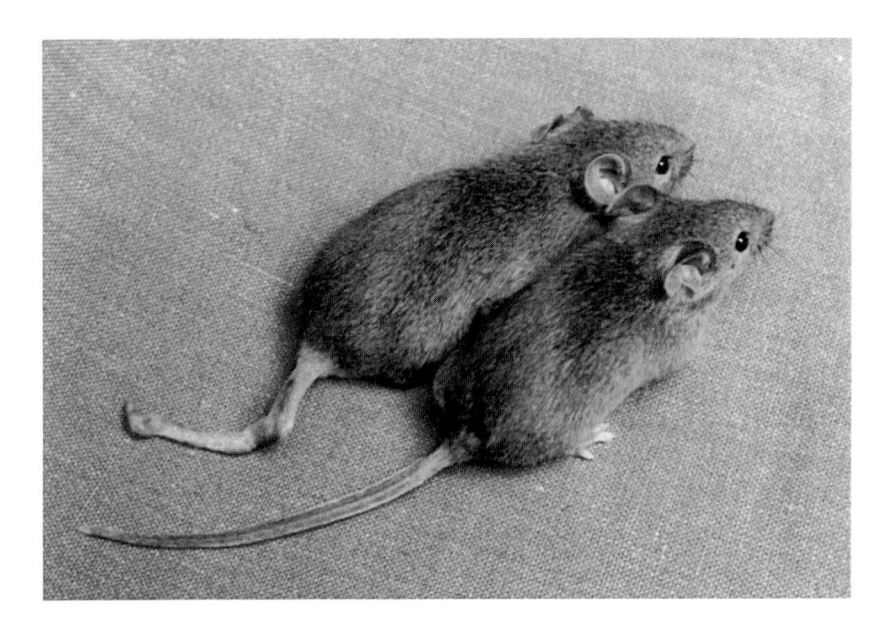

Figure 3-2: These mice are identical clones (the same genetically) but the normal mouse on the right can be contrasted with the mouse with a high degree of DNA methylation on the left. Source: Emma Whitelaw

It was natural that parents with known genetic disorders would want to protect their children from inheriting often crippling diseases. Research sought to help individuals with mitochondrial diseases passed down from their mothers and grandmothers. One technique that was developed was known as three-parent in vitro fertilization which involved taking genes from the mother, father and another female, allowing the replacement of the faulty mitochondria. The importance of this development was underlined by Lisa Jardine, the Chair of the Human Fertilization and Embryology Authority in the UK when she said that:

> *"This is uncharted territory. If this is allowed...it has consequences in perpetuity....it has the potential to alter the nature of society to all eternity."*

There were a number of different approaches to treating mitochondrial diseases including pronuclear that involved the swap of DNA between two fertilized eggs, and maternal spindle transfer that involved the transfer of genetic material between the mother's egg and a donor egg before fertilization. Ethicists and advocates for the handicapped expressed their fears that these and related procedures were only a first step to eventually conducting complete genetic redesign of offspring in attempts to create a "master race" of genetically superior humans. (Kelland 2012)

In September 2013, a company located in Mountain View California with the rather playful name of 23andMe, was granted a US patent for a technique to predict the various traits of an infant based on the DNA of the parents. The method could be used to screen sperm and ova (eggs) that were to be employed for artificial fertilization for such common traits such as gender, size, eye colour, disease susceptibility and much, much more. Critics immediately slammed the company for giving parents that could afford it, the capability of creating Just-On-Order-Babies. There was also the unsavoury possibility that genetic disorders such as Parkinson's would become concentrated among the segments of the population that could not afford genetic screening. (Callaway 2013)

Scientists in Britain and Japan developed a way to significantly accelerate the evolutionary process. The method was to find naturally occurring sections of DNA that were known to confer specific qualities in an organism. Using this information, it was possible to remove these strands and then splice them into another organism. (Vancouver Sun 2012) While the original purpose of the research was to develop favourable characteristics in plants, it would only be a matter of time before these techniques were used to transfer favourable characteristics to selected human beings. Will wealthy parents be able to resist the temptation to confer superior abilities to their children and grandchildren? Will future authoritarian governments not be willing to create enhanced athletes to promote the superiority of their political systems? These were only a few of the profound ethical questions that will have to be addressed very soon, before this research progresses to subsequent stages and eventually to full commercialization of the technology.

An app became available for the iPhone and iPad called Proloquo2go designed for individuals who had difficulties communicating, such as autistic children or elderly stroke patients. All the user had to do was to point to a series of pictographs representing various words and the program used a synthetic voice to speak the chosen words out loud. This app was one of the first of a whole host of different medical apps that were being made available for download. As another example, the Centre for Global eHealth Innovation in Toronto developed "Bant," an app designed for use by Type-I diabetics. All the patients had to do was to prick themselves with an electronic glucometer that was equipped to wirelessly communicate with the Bant app. The software was designed to warn patients when their blood sugar patterns were abnormal or irregular. Hundreds of other highly sophisticated apps were under development to monitor everything from heart rate and blood pressure to the regularity of bowel movements or sexual response.

A bioengineer named Aydogan Ozcan developed an app that could convert an ordinary cell phone camera into a laboratory grade, blood-cell analyzer that eliminated the need to send samples away for analysis. The blood sample was simply placed onto a glass slide and positioned in front of the camera. It could then be screened for HIV, thalesemia, sickle-cell anaemia, malaria or a host of other blood borne disorders. The modern smart phone that was equipped with computational and graphical analysis power that would have cost millions and weighed tonnes a generation earlier, was now configurable as a powerful diagnostic tool weighing about the same as a paperback book. Eventually, a phone that fit into your pocket would be able to conduct a complete genomic analysis of you based on a single blood or skin sample. (Freeman 2012)

The company, Claros Diagnostics, working in collaboration with researchers at Columbia University developed a device that was essentially a "blood testing lab in a box." With a volume of less than 40 cubic inches, the device consisted of a microprocessor, micro-pump, photo-detectors, LEDs, and a 9 Volt battery. Replacing laboratory equipment worth more than $100,000, the $100 device was able to instantaneously test blood samples for HIV and a whole host of other STDs. The operation of the device was enabled by advances in the science of microfluidics, the manipulation of tiny volumes of fluid and improvements in antigen biochemistry that allowed a highly portable device to detect ever increasing numbers of diseases by testing ever decreasing volumes of sampled blood. (Angelle 2010)

A small hand held device called PreVu was designed to warn people if their risk of heart disease was normal, marginal or high. The first generation could not distinguish between the different types of cholesterol nor could it give an actual numerical measurement of the exact amount of the substance in the blood, but it could issue a warning that a person was "strongly encouraged" to go see their physician. The device operated by measuring the amount of cholesterol in the skin. But since the correlation between the amounts of cholesterol in the skin and blood was not perfect, the machine could not generate an exact number, only a general warning. Future generations of such devices will harness a variety of testing methods to generate increasingly accurate numerical measurements of a number of important parameters including blood sugar, hormone and cholesterol levels.

A middle manager in a successful mid-size company noticed a welt on his skin. He thought it was a simple acne eruption, but it was not healing. In the past, he would either have ignored it and then waited until it got really bad, or he would have had to take the initiative to take time off from his busy work schedule to go and see his dermatologist. He did not want to do this, so he decided to use a third alternative that had recently been made available to him. He removed an electronic device that looked like a black box about the size of a pocket calculator from his pocket. He turned on the device and swept it once over the suspicious area.

The device used an infrared laser focused with a liquid lens capable of refocusing more than 30 times per second while taking thousands of images of cells located in the welt and in the area around it, to a depth of 2 mm. The device then linked to a cloud-based supercomputer that was able to join together the series of pictures, creating a 3D virtual biopsy of the welt that allowed it to be viewed from any orientation. In the first generation of the system, the virtual biopsy was examined by human professionals to determine if the welt was malignant or not. However, in future versions, the analysis of the biopsy will be done automatically and the results sent to the patient's GP so that a referral could be issued immediately to see an oncologist.

Doctors already had access to apps such as Epocraes, Medscape and Micromedex that allowed them to calculate exact doses for a wide variety of drugs based on the age and weight of the patient. These apps ushered in a new and exciting age of Just-On-Order-Medicine where each medical professional would at all times have access to the results of the latest medical research news, clinical warnings, pharmaceutical information and medical calculations. Of even greater potential impact, advanced diagnosis apps allowed physicians to verbally describe the symptoms experienced by a patient and the software would then list a number of the most likely causes, along with the suggested prescriptions. Such apps would not be replacing medical professionals any time soon, but they could help by reducing the incidence of errors to which doctors were frequently prone such as "confirmation bias" or "anchoring."

An example of the latter was a patient who presented with all of the symptoms of Crones' Disease, so the physician became anchored on that diagnosis by seeking more symptoms that confirmed the diagnosis, while subconsciously ignoring those that did not. In fact, a MRI confirmed that the patient actually had intestinal cancer, and the delay that occurred because of the initial misdiagnosis meant that the establishment of the proper diagnosis happened too late to save the patient.

The problem in developing such apps was that because of the great variability between humans, no two individuals reacted identically to the same disease. This was one of the reasons why a impending heart attack was so difficult to diagnose in women, because it could be manifested by so many different symptoms, ranging from chest or breast pain, to nausea and headache. Consequently, for the foreseeable future, the most powerful tool that practicing physicians had at their disposal remained a solid education, years of clinical experience and a highly developed sense of judgment. (Aw 2011)

Using the rapidly evolving science of Optogenetics, it became possible to design Just-On-Order-Mental-Health-Treatments for various psychiatric disorders. Developed at Stanford University's Clark Center, it was based on the clinical observation that light could be used to control the neural activity of the brain. After injecting a specific light sensitive protein, optogenetics involved switching brain cells on or off by exposing the surface of the brain to targeted green, yellow or blue flashes of light. It had long been known that certain microbes were capable of making light-sensitive proteins such as bacteriorhodopsin (that responded to green light) and opsin halorhodopsin (yellow light). After a significant amount of experimentation, the refined technique involved the following steps:

1. A gene that made cells light sensitive was isolated from an organism

2. A DNA strand was added that allowed the gene to bond with a particular neuron

3. The tagged gene was inserted into a harmless virus that was then injected into the patient's brain

4. The targeted neurons had the gene injected into them by the virus and they soon expressed the light-sensitive protein on their surfaces

5. Using an optical fibre, the neurons were illuminated by the intense burst of coloured light

6. The proteins in the neurons responded to the light, turning on or off the neurons in the targeted centre of the brain.

This approach was successfully used to stop cocaine addiction in mice. However, the research opened the door to the possibility of influencing the brain and consequently controlling behaviour at the cellular level. The possibility existed of creating trans-genetic humans who had the majority of their behaviours moderated and controlled by optogenetics. One thing about the technology was certain. Would the analysis of the ethical and moral implications of the technology be able to keep up with the progress of the technology itself? And if not, was this the long sought Dictator's Dream?

Neural interconnections varied significantly from individual to individual. A kind of brain technology known as Diffusion Tensor Imaging (DTI) permitted the construction of detailed maps of these individual neural interconnections. Using these maps, abnormalities could be located, identified and then a technology such as Transcranial Magnetic Stimulation (TMS) could be used to irradiate the affected centres of the brain with magnetic pulses that were powerful enough to reconfigure the faulty neural interconnections. TMS has already been used to treat mental disorders such as Parkinson's, bipolar disorder, migraines and depression. (Barth 2012) TMS could also be used to selectively erase memories from specific regions of the brain. What possible applications could the *powers that be* develop for this technology that would permit each individual to have Just-On-Order-Memories?

Just-On-Order-Medical Sensors & Devices

Radio-Frequency-Identification (RFID) has been used for numerous applications such as product tracking, inventory control and security, but a novel application was in the real-time, *in-situ* (inside the body) monitoring of healing bone fractures by Osteo-synthesis. The entire system consisted of a metallic implant mounted on the broken bone, microcontroller unit, strain gauge, RFID chip and an external signal reader. The advantage of the system was it required no internal power source, since all of the required power to operate the system was provided by the signal emitted by the reader unit, typically held closer than eight centimeters to the surface of the skin, to effectively communicate with the implanted transponder. The system successfully monitored the healing of fractures over time without the patient having to be repeatedly exposed to the radiation emitted by X-rays or CT scans. In the future, the system will be expanded to provide real-time monitoring of other biological implants such arterial stints, heart valves and hip implants. The technology will be gradually improved so that readings could be made from increasingly greater distances. (Dirjish 2010)

An innocent looking hand-held device was developed by Andre Marziali, a physics professor at the University of British Columbia, which could make a rapid identification of virus DNA. The most difficult task was separating the background human DNA from the genome of the infectious agent. The patient supplied either a blood sample or a throat swab, and these would then be analysed by the device. This device was developed from the experiences obtained in the design of Aurora, a machine that could detect, extract and analyse DNA from heavily contaminated samples. The key technological advance used in both of these devices was an electric field generator employed to untangle and straighten the normally tightly coiled DNA molecules. After a number of technical improvements, the machine will be used to identify the DNA of potential pandemic agents before they could start killing off human beings, or the DNA of cancer cells such as those found in malignant pancreatic tumors. (Schwartz 2013)

Another rapidly maturing technology was Micro-Electromechanical-Systems (MEMS). Initial applications involved health monitoring of post-surgical patients and athletes. This became possible because of the development of increasingly miniaturized accelerometers, gyroscopes, GPS systems and inertial measurement units. Every movement made by a completive athlete during an event could be monitored in real time and recorded for later analysis. When coupled with an extremely Low Power Bluetooth (LPB) wireless device, many bodily parameters associated with health and wellness could be monitored for extended periods of time. Apps were available to link to LPBs so that smart phones could be transformed into effective and highly portable monitoring devices. MEMS based devices were developed to monitor heart rate, blood pressure, aneurisms, brain waves and much, much more.

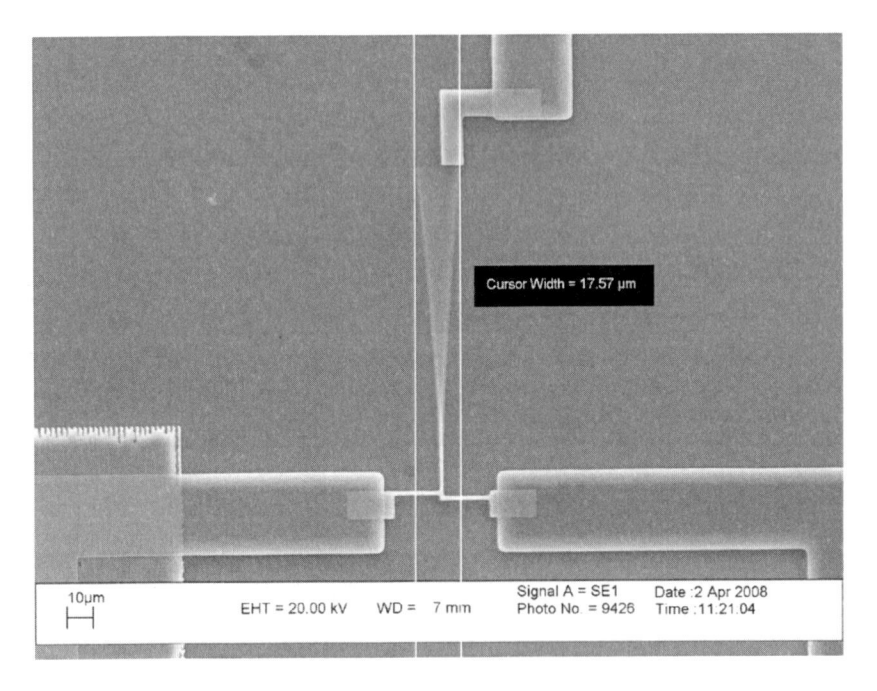

Figure 3-3: Example of an MEMS sensor consisting of a vibrating cantilever beam. The whole device is less than 0.02 mm in size. Source: Pcfleto1

The field of Microfluidics involved the use of MEMS actuators to dispense fluids. As an example, a new technology insulin pump could be placed onto a disposable skin patch to provide a continuous supply of insulin for diabetics. The University of British Columbia developed a MEMS-based drug delivery system that was implanted behind the eye so that it could release drugs on-demand to treat diabetes-related vision loss. The key component of this system was a reservoir covered with a magnetic membrane. Exposing the device to an external magnetic field caused the membrane to deform and to eject a measured quantity of drug into the system. Leading edge research is investigating the use of MEMS devices, based on stacks of silicon wafers with micro-etched channels, as substitutes for filtering organs such as kidneys and livers. (Allan 2012)

Rapid advances in computer technologies opened up the possibility of Just-On-Order-Mobility for thousands of para and quadriplegic human beings. The technique involved implanting multiple electrodes into critical locations of the patient's brain, in some cases consisting of less than fifty neural cells in the primary motor cortex. The signals detected in the brain were then interpreted by the computer and then used to issue electrical signals to the nerves of the major mobility muscles. Simultaneously, electrical signals from the fine sensory nerves in the extremities where taken by electrodes and fed into the computer, where the signals were interpreted and then converted into signals that could be conducted into the brain's primary somatosensory cortex, allowing the patient to get feedback on various parameters, such as the texture of an object. The achievement for the first time of the combination of movement with sensory feedback was the breakthrough that patients with mobility issues had been waiting for to offer them the chance to lead more fulfilling, independent lives.

Nature has provided the inspiration for many human innovations. The various properties of the Atlantic Sea Lamprey served as inspiration for the development of a swimming robot known as the "Cyberplasm" that was designed to be highly sensitive to the environment in which it was moving. Using leading edge nano-electronics and biomimetics, it was able to sense the nature of its environment by detecting light, temperature and chemical concentrations just like the sea creature after which it was modelled. Eventually the goal was to miniaturize the device to a size where it could easily swim through the human circulatory system "sniffing" out diseases that were only in the very earliest stages of their development, improving the odds of recovery. The next stage of development, after the dimensions are reduced as far as possible, will be the large scale production of such exploratory robots using 3D making technologies.

A number of innovative researchers found that the growing power of social media could be harnessed for a number of medical applications. One developing application was to detect and monitor global pandemics. An epidemiologist named Dr. John Brownstein created a web site called HealthMap, a site and mobile app that was designed to constantly explore the Internet, searching blogs, news and social networking sites for evidence of the early stages of symptoms of respiratory infection that could indicate the beginning of a pandemic. These systems were capable of pre-emptively evaluating the seriousness of every new disease outbreak, allowing laboratories to prepare an immediate response, before the infection could become widespread. While the system was certainly not foolproof, it was definitely a significant advance over traditional government-run infection detection systems. (Weeks 2011)

Just-On-Order-Medical-Research

For generations, medical research was conducted according to what was known as the "Traditional Model," The required steps in this model always consisted of the following:

- The formulation of a hypothesis

- Design of a study and recruitment of participants

- Conducting the study (or studies) to test the hypothesis

- Data aggregation and analysis

- Writing the paper and submission to a medical research journal

- Acceptance of the paper

- Publication of the research results

The time taken from the formulation of the first hypothesis to the final publication could exceed six years and cost millions. The advent of the Internet and social networking technologies allowed the development of a new research model. This new approach to medical research involved the following alternative steps:

- Design of the survey to gather statistical information on symptoms

- Using social media to recruit thousands or even millions of research subjects

- Aggregation of data by automated sensor analysis of participants and completion of surveys

- Data reduction and analysis by using a database query based on all the subjects of the study

- Publication of the results on-line

The total time elapsed with the new model would be less than eight months. Millions more patients could be helped by the new treatment. This demonstrated the power of the new on-line technologies. (Goetz 2010)

Conclusion

Rapid, inexpensive and accurate DNA sequencing will allow geneticists to determine the probability that a particular individual will develop certain diseases. It will be possible for those at greatest risk to begin therapies for disorders that they had not yet developed symptoms for. With gigabytes of genomic data generated for each individual, it will be possible to develop drug treatments specially tailored for each person.

Continuous monitoring will allow people to maintain their own health not only by treating their known existing illnesses, but by dealing with their susceptibilities to future illnesses. The challenges will include the difficulty in maintaining privacy in an environment where everything about each person is known right down to the molecular level. There is no guarantee that this knowledge will be used ethically. For example, once all of genes for aggression and deviant behaviour have been identified (assuming such genes exist), will social engineers combine forces with genetic engineers to start tinkering with an individual's genome to create a "better" citizen?

Using molecular 3D printers and other technologies, medicine will soon have the ability to construct from scratch, an entire synthetic DNA molecule, and will be able to conduct the systematic analysis of the function(s) of every base pair in the DNA molecule, of every gene, and of every protein encoded by each gene. The eventual result will be the construction of Just-On-Order-Organisms designed to serve useful functions in environmental engineering, biofuels, chemical engineering, vaccines and other pharmaceuticals. Extinct creatures from the past could be re-created much more efficiently than by using the primitive techniques suggested in sci-fi films such as Jurassic Park. Brand new, highly beneficial organisms, that have never before existed on Earth, could also be designed with equal ease. But such research could also be used to develop new bacteriological warfare agents for which no vaccines could be developed. Once again, it is not the technology that is evil, but the ethics of the user and of the application.

The capabilities of 3D printers to manufacture replacement tissues and organs made from a person's own stem cells will continue to improve. While initially organs will be grown outside the body and then will require a surgeon to implant, eventually replacement organs will be grown inside the body and will gradually take over the function of the damaged organ. The world is already facing a dramatic and sustained population increase in the elderly, but inexpensive organ production and replacement would mean that increasing numbers of people would be capable of living to advanced old age, likely straining social support networks to the breaking point.

How far the 3D printing of tissues will go was explored at a World Future Society conference held in Chicago in the summer of 2013. A Russian researcher stunned the audience that consisted mostly of professional futurists, when he said that within the next ten years it would be possible to print out an entire human being, from big toe to head hair and everything in between. How the brain of such a "mass produced" being would be filled with all of the information necessary for proper biological and social functioning was not addressed.

In just about every area of medicine, technologies were advancing at a breath taking pace. With an ever deepening understanding of how to repair damage to the nerves of the central nervous system, it appeared that head transplants would soon be possible. Imagine a scenario where a seventy-one year old man was suffering from terminal cancer that had spread throughout his body, but had not affected any of the tissues above his neck. A twenty-two year old suffered a fatal head injury in a serious high-speed motorcycle accident and was declared brain dead in the ER. But he had suffered only minimal damages to his body. It will soon be possible to remove the healthy head from the body of the seventy-one year old, and place it onto the cancer free body of the twenty-two year old.

Now let's say, hypothetically speaking, a Jewish head was placed onto a Muslim body. What religion will the resulting hybrid person be? Would it be acceptable to place a female head onto a male body resulting in an instant sex change?

This procedure and many similar to it will be rife with ethical issues, systems effects and unintended consequences that policy makers of the future will have to grapple with. And this is only the tip of the ethical iceberg, because as mentioned previously, beginning with human stem cells, 3D bioprinting will soon be able to produce entirely new bodies and eventually, even new heads.

Some of the most significant technological trends associated with Just-On-Order-Medical-Monitoring were examined. In principle, it was already possible to assign every person on earth and every organ within each of those people their own, or its own IP address. These organs will be able to communicate with what could be called a massive e-health network, designed to constantly monitor the health of everything and everybody. Implanted sensors will constantly take readings of various parameters such a blood pressure, heart rate, pH etc. and then relay these in real-time to internet linked cell phones and car computers. The bottom line was that a significant amount of the preventative care now carried out by health care professionals will gradually be shifted to auto-mated sensors and networked software designed to operate as a global body-area-network. The technology that enabled the establishment of this network was the presence of more than six billion cell phones linked to networks operating even in the remotest parts of the globe. The Body Area Network will link to Home Area Networks that will automatically link with Urban Area Networks, and finally to a Global Knowledge Network consisting of autonomously operating sensors operated by intelligent software systems.

In all of these discussions, one very rapidly developing area of medi-cine was not even been mentioned yet. The technology of *Tele-Medicine* was capable of making the talents of the very best surgeons in the world available instantly, anytime and anywhere that they were needed. At one end of the system, the surgeon held the handles of their favorite surgical instruments in his/her hands while looking at the patient through a 3D virtual reality headset. The movements of the instruments were con-verted into digital signals that were transmitted over the Internet to a device that controlled the working end of the surgical instruments being manipulated by the surgeon. It was these remotely controlled blades and clamps that performed the actual surgery on the patient. Though the surgeon and the patient may be separated by half a world, to the doctor, it felt for all intents and purposes as if the patient was on the table in front of them. This technology will in principle allow a patient in any remote jungle village to receive the same quality of medical care as a patient in the most modern clinic in the middle of great urban centre in the developed world.

The next steps to developing this technology will be to use 3D print-ers to manufacture all of the components necessary to construct a complete tele-medicine facility, and then gradually letting autonomous robots take over many of the surgeries now being done by human doctors.

All of these are exciting components of your JOOM destiny.....

Web Resources

3D Bio-Printing: http://www.youtube.com/watch?v=9D749wZSlb0

3D Printing of New Tissues: http://www.youtube.com/watch?v=5SWw_qM6_8I

3D Printing of Human Organs: http://www.youtube.com/watch?v=nEPz-OLxosA

Bio Nano Technology: http://www.youtube.com/watch?v=sjV7NNwm1GU

Printing Body Parts: http://www.youtube.com/watch?v=Ly5h252TRfM

Printing Human Organs: http://www.youtube.com/watch?v=4nqw1yjyKEs

Printing a Human Heart:
http://www.policymic.com/articles/83177/in-amazing-feat-of-science-surgeons-save-child-s-life-by-3-d-printing-a-new-heart

Photo Credits

Figure 3 1: http://commons.wikimedia.org/wiki/File.BrainGate.jpg

Figure 3-2: http://commons.wikimedia.org/wiki/File:Cloned_mice_with_different_DNA_methylation.png

Figure 3-3: http://commons.wikimedia.org/wiki/File:MEMS_Microcantilever_in_Resonance.png

References

Abraham, C. (2008) "Genetic Find Puts Scientists a Step Closer to Generating Replacement Organs," *Globe and Mail*, August 7, p. 1

Adafruit (2013) "Fascinating & Frightening Ways 3D Bioprinting is the Next Big Thing in Medicine & Science," May 2, https://www.adafruit.com/blog/2013/05/02/fascinating-and-frightening-ways-3d-bioprinting-is-the-next-big-thing-in-medicine-and-science-3dthursday/

Allan, R. (2012) "System-Level Applications Make MEMS Ubiquitous," *Electronic Design*, January 12, p.61

Angelle, A. (2010) "How It Works: The Lab That Fits in Your Hand," *Popular Science*, April, pp. 48 - 49

Anthes, E. (2011) "Stem-Cell Therapy Works Wonders for Race Horses: Are Human Treatments Next?" *Popular Science*, June, p. 33

Aslam, T. (2012) "Bioengineers Use 3D Printer to Create Human Organs," *The Daily Pennsylvanian*, September 24

Aw, J. (2011) "Paging Dr. Smartphone: How Medical Apps Are Changing Diagnoses and Treatments," *National Post*, December 8

Barth, A. (2012) "Controlling Brains with a Flick of a Light Switch," *Discover*, September, p. 37

Beasley, D. (2012) "Researchers Gauge Impact of Discovery," Health Section, Globe Life, *Globe and Mail*, Aug. 12, p. L5

Begley, S. (2010) "Breasts are Just the Beginning," *Wired*, Nov. p. 150-188

Bioscience World (2012) "Auto Genomics, Revolutionizing Molecular Testing," Laboratory Focus, *Bioscience World*, March, pp. 12-15

Bohan, S. (2011) "Gene Therapy Shows Promise," *Vancouver Sun*, October, 11

Brown, E. (2012) "Scientists Work Toward Creating a Living Hard Drive," *Vancouver Sun*, June 2.

Calamia, J. (2011) "Hit 'Print,' Make Blood Vessels," *Discover*, July-August

Catalyst (2012) "Organ Bioprinting," October 25, http://www.abc.net.au/catalyst/stories/3618385.htm

Cheng, M. (2013) "Doctors Grow New Vein with Girl's Own Stem Cells," *Toronto Star*, June 13

Callaway, E. (2013) "Personal-Genetics Firm Denies Pursuit of 'Designer Babies,' "*Health News*, Scientific American Web Site, October 2.

Carman (2011) "BC Doctors Link Single Mutation to Cancers," *Vancouver Sun*, December 22, p. 1, p. 13

Chant, I. (2012) "Artificial Eggs and Artificial Sperm Produce Real, Adorable Baby Mice," *Geek System*, October 5, http://www.geekosystem.com/artificial-eggs-sperm-mice/

Chivers, T. (2013) "Artificial Body Parts Take Big Strides," *Vancouver Sun*, Feb. 23.

Cohen, S., Leor, J. (2004) "Rebuilding Broken Hearts," *Scientific American*, Nov., pp. 45-51

Dirjish, M. (2010) "RFID Technology Monitors Bone Fractures As They Heal," *Electronic Design Magazine*, June 10, p. 67

Emory, A. (2002) "No Bones About It – Rapid Prototyping – The Future of Bioceramic Implants," *Design Engineering*, August/September, p. 32

FAS (2001) "Mousepox Case Study – Unintended Experimental Results in Transgenic Research," Federation of American Scientists, http://www.fas.org/biosecurity/education/dualuse/FAS_Jackson/index.html

Faulker-Jones, A. et al. (2013) "Development of a Valve-Based Cell Printer for the Formation of Human Embryonic Stem Cell Spheriod Aggregates," *Biofabrication* 5 015013 DOI: 10.1088/1758-5082/5/1/015013

Fayerman, P. (2012a) "Researchers Unlock Deadly Disease's Code," *Vancouver Sun*, April 5, p. A11.

Fayerman, P. (2012b) "Treatments Tailored to Tumours," *Vancouver Sun*, January 31

Freeman, D. (2012) "A Doctor in Your Pocket," *Discover*, September, p. 24

Freeman, F. (1960) "Cyborg," Painting in *Life Magazine*, July 11, http://cyberneticzoo.com/tag/cybernetic-organism/

Globe & Mail (2012) "New Approaches Aim to Detect Potentially Aggressive Breast Cancers," Breast Cancer, *Globe and Mail*, September 28, p. BC2

Goetz, T. (2010) "Sergey Brin's Search for a Parkinson's Cure," *Wired*, June 22, p. 107

Gray, R. (2012a) "Human Brain Cells Created from Skin," http://www.telegraph.co.uk/science/science-news/9076852/Human-brain-cells-created-from-skin.html

Gray, R. (2012b) "Vaccine Has Victim's Own Body Attack Cancer," *Vancouver Sun*, April 09.

Graziano, A. (2014) "Bioprint," https://www.youtube.com/playlist?list=PL9-reVVZEI9SOrO39wDVlsqbJv4VzLnS7

Halley, D. (2009) "Growing Organs in the Lab," Singularity *HUB*, June, http://singularityhub.com/2009/06/08/growing-organs-in-the-lab/

Kelland, K. (2012) "Three-Parent Embryos Unnerve Ethicists," *Globe and Mail*, September 18, p. L9

Kladko, B. (2011) "UBC Researchers Create More Powerful "Lab-On-A-Chip" for genetic analysis," *UBC Public Affairs*, July 26

Lauerman, J. (2008) "Fast DNA Mapping Will Change Health Care," *Vancouver Sun*, November 21, p. B3

Lauran, N. (2013) "Challenging the Transplant System," *National Post*, June 8, 2013, p. A15

Lenzer, J. (2009) "The Super Cell," *Discover*, Nov., p. 31

Macrae, F. (2009) "Ethical Storm Flares as British Scientists Create Artificial Sperm from Human Stem Cells," *Daily Mail*, July 8

Marchione, M. (2012) "New Medicine Helps Troops," *Vancouver Sun*, September 11, p. B5

Moyer, M.W. (2011) "Organs-on-a-Chip," *Scientific American*, March, p. 19

Nano Lett., (2013) "3D Printed Ears," 13 (6), pp 2634-2639 DOI: 10.1021/nl4007744, May 1.

Natalizio, D. (2008) "IT Makes Personalized Medicine a Reality," *Biotechnology Focus*, January, p. 22

Parenteau, N. (1999) "Skin: The First Tissue-Engineering Products," *Scientific American*, April, pp. 83-89.

Pilger, R. (2012) "One Breath Away From Diagnosing Disease," *Globe and Mail*, February 4, p. A15.

Piore, A. (2012) "Big Idea: Growing Livers in Lymph Nodes," *Discover*, March, P. 10

Power, A., Martin, V., Kaern, M. (2012) "Intelligent Design: Synthetic Biology and the Evolution of Biotechnology," *Laboratory Focus*, May, p.10

Richmond, S. (2012) "3D Printer Builds New Jaw Bone for Transplant," *The Telegraph*, Feb. 7, http://www.telegraph.co.uk/technology/news/9066721/3D-printer-builds-new-jaw-bone-for-transplant.html

Rockel, N. (2011) "Fantastic Voyage Redux: Smarter Implants are Coming," Report on Business, *Globe and Mail*, November 2, p. B12

Schaffer, A. (2008) "Creating a Heart," *Technology Review*, May/June, p. 88.

Schwartz, A. (2013) "The 16-Year-old Who Created a Cheap, Accurate Cancer Sensor is Now Building a Tricorder with Other Genius Kids," *Fast Company*, February 26

Sheridan, K. (2010) "Stem-Cell Transplant Keeps Patient HIV-Free for Three Years," *Vancouver Sun*, December 16

Spears, T. (2012) "Canada Develops First Bedside DNA Test," *Vancouver Sun*, Sept. 11

Stone, A. (2009) "Cyborg Therapy," *Forbes Magazine*, Health Section, February 16, p. 68

Stroh, M. (2003) "Print Me a Pancreas, Please," *Popular Science*, May, p. 51

Sullivan, J. (2013) "Printable 'Bionic' Ear Melds Electronics and Biology," Princeton University Engineering School, May 1, (http://www.eurekalert.org/pub_releases/2013-05/pues-pe050113.php)

Tam, D. (2013) "That Makes Two of Us: How Bioengineers are using 3D printing to create Body Parts." *South China Morning Post*, June 29.

Taylor, P. (2012) "Scientists Find a Master Control Gene for Blood Stem Cells," *Globe and Mail*, Nov. 8

Ubelacker, S. (2012) "Stem Cells Restore Hearing in Gerbils," *National Post*, December 9, p. A10

Vancouver Sun (2011) "Girl Undergoes Six-Organ Transplant," Feb. 6.

Vancouver Sun (2012) "Technique Could Boost Crop Yields," January, 23, p. B4

Weeks, C. (2011) "Social Media Could Help Detect Pandemics," *Globe and Mail*, June 27

Weeks, C. (2012) "Why a DNA Test May Not Save Your Life," Globe Focus, *Globe & Mail*, Feb. 11, 2012, p. F5.

Zapana, V. (2010) "Made-To-Order Lungs," *Popular Science*, October, p. 30

Chapter 4
The Third JOOM Revolution
Just-On-Order-Media

Introduction

The 19th Century provided many of the technologies such as mass print media, sound recording and radio that allowed the various media industries to prosper in the 20th century. But the industry was always highly sensitive to changes in technology. For example, the introduction of television devastated both the radio and movie theatre businesses. At one time there were three very powerful radio networks, but today almost no one even remembers that they ever existed. With the rapid penetration of TV into the North American home, it was estimated that more than 30,000 neighbourhood drive-ins and movie theaters were forced to close. However, the advent of almost universal television ownership in the second half of the 20th century gave the industry a degree of power and influence over people's lives that could never have been imagined in any other historical period. By the end of century when they achieved the peak of their power, the policy makers of the various media industries felt that their hold on the population was as secure as ever. Advertising moguls, the movie and music studios, as well as the three major television networks all shared the same view and had the utmost confidence that their grasp on the content and technologies of media would be secure for the foreseeable future.

Then came the Internet and the World Wide Web. One of the purposes of the traditional media was to expose the viewing audience to new ideas, or at least to the ideas that were considered acceptable by the media executives. The on-line media was able to accomplish this task to a much greater extent since it was available 24/7 anywhere in the world where there happened to be an internet connection. One of the first of these next generation media Ideagoras was FORA.tv. It allowed anyone to watch idea-rich content on their TVs, smart phones or on their tablet computers. The strength of this new media was that the video content could be shared along with the commentary and feedback provided by previous viewers. More recently, *TED TALKS* (www.tedtalks.com) brought together, on one site, researchers, scientists and plain interesting people to present global paradigm shifting ideas to a worldwide audience. While traditional media encouraged viewers to watch passively, the new media wanted to stimulate conversation and debate. Furthermore, the Web encouraged experimentation since everything could be set up so quickly. If a concept did not work well, it could be instantly changed and tested again, or taken down completely, all in less time than it traditionally took a media mogul to eat a caviar covered cracker.

The Advertising Industry

"The name of the game is to be relevant." Kent Anderson, President of Macys.com

According to documentary film maker, *Amr Salama*:

"Businesses now really need to understand something that governments and dictators, didn't understand. They didn't understand that someday they'd be busted. Anything you do will be known. Social media gonna get you, and if you're lying, we're gonna know....Yesterday it was the government, tomorrow it's going to be the corporation." (Houpt 2011)

By the opening years of the 21st century, the era of passive TV watching was rapidly drawing to a close. Young people by the millions were abandoning TV and embracing on-line entertainment. With the arrival of the relatively inexpensive tablet computer, TV watchers were increasingly using the connected devices in real time to research the programs they were watching. Worried that they might turn off the tube entirely and move totally onto the Web, the networks responded by producing apps that offered significant volumes of additional "enhanced" content.

The next step taken to keep users bonded to their TV sets was to enable the tablets themselves to communicate with Wi-Fi enabled TVs. The purpose of the apps was to identify what program the user was watching using an audio recognition algorithm and then to create a virtual library of related materials taken from many different Internet sources. It was only one more step beyond this to customize all of the information that was selected to flow towards each individual viewer. As an example, Microsoft released the Xbox *SmartGlass* app that allowed videogame developers to create split-screen experiences that were specifically designed for each user. The advertising industry was forced to adapt to these rapidly evolving technologies, or perish along with so many other industries that simply failed to adapt.

It was certainly a gross understatement to point out that the advent of the Internet and supporting technologies led to dramatic changes in the way business was conducted. This was especially true for the advertising industry. The changes were so profound for this sector that it would be fair to say that the history of the industry could be divided into two ages, BI and AI, or *Before the Internet and After*. There were four major trends leading these changes; The shift from a consumer society to an information society; The transformation from a mass society comprised of identifiable groups with specific characteristics to a society of individuals; The evolution from a more formal society to one that could be considered to be informal and even casual, and; The gradual transformation from a class-based society to one based on inter-personal, inter-group and inter-organizational relationships. (Seguela 2010)

The Internet permitted advertising to take directions never before imagined. Designer Ralph Lauren developed an iPod App capable of displaying high-definition 3D models of all of the possible combinations of colour, shapes and inscriptions that were available for its best selling purse known as a Ricky Bag. Using newly developed scanning technologies, it became possible to record all of the body measurements of a particular client and to show in a full 3D rendering how they would look in a particular clothing outfit with different combinations of accessories.

The rulers of the now defunct Soviet Empire expressed great concern that their captive populations completely tuned out most of the all encompassing propaganda that was supposed to inspire them and keep them filled with revolutionary fervour. Similarly, throughout the consumer societies of the West, faced with an onslaught of advertising from every direction, people were becoming irritated, defensive and bored. Increasingly they were tuning out the old media, turning on to new technologies and dropping out as members of a mass media audience.

The sector was discovering that to keep working, advertising had to leave behind its tried and tested 20th Century mass messages and embrace a much more difficult strategy...to develop a new approach that used technology to create custom designed messages that were targeted laser-like at individual consumers.

In the 21st Century, consumers were to be increasingly found within the protective comfort of specialized groups of all types. A search of the Internet could uncover, for example, a special group for gay horticulturalists that grew red roses. Only with the invention of the Internet and the World Wide Web could geographically disparate individuals with similar interests effortlessly get together to share experiences and insights.

Rather than being passive consumers of the messages issued by advertisers, the new consumer would gladly take on a much more proactive role. They wanted to click on the ad that they were interested in on their ultra-thin screen, smart TV. They wanted to extract the information that they wanted, when they were ready to receive it. Interactivity became so universal that it led to the invention of a new term to describe the involved, non-passive, proactive consumer – *the prosumer.*

The prosumer would no longer passively sit and watch videos, but interacted with them on multiple levels. For example, information on all of the products appearing in the video would be instantly available. A decision could be made while the story was still developing, if one desired to make a purchase. In many cases, rather than having to wait for delivery, the prosumer could choose to download the 3D design for the product desired and it could be manufactured then and there in the person's living room using a personal 3D printer. Even if they did not choose to purchase the products that they saw being used in the films and programs that they are watching, prosumers would always have the option to find out much more about them because each item presented would come with an instantly accessible link that could be followed to find out everything that could be known about a particular item.

Until the beginning of the 21st Century, all advertising was a one-way conversation from the seller to the potential buyer. Consumers were never asked for their opinions and were not solicited for their input. In fact, the traditional advertising industry could have been considered to be little more than a producer of propaganda. It was assumed (not without justification) that consumers were ignorant, immature and one-dimensional. But now with various social media tools, consumers were linked together in ad-hoc knowledge/educational networks where they could find out as much as they wanted about anything they happened to be interested in.

In the hope of unifying the world, visionaries of the past developed a simple language known as Esperanto, in the vain hope that people around the globe would take the time to learn it and then be able to communicate with anyone else on earth. In actual fact, the Internet provided people everywhere with a de-facto digital Esperanto. With the instantaneous translation of languages, the age of truly universal communication finally dawned where in principle every human mind could interact with every other mind on the planet, transcending language, race, culture, education, religion, class, geography and politics. The full implications and impacts of this historically unique situation remain to be seen, but they are likely to be broad, profound and sustained.

During the Christmas Season of 2012, prosumers experienced a barrage of e-mails targeting their smart phones. *"If you buy today, we pay the shipping!"* said one. *"During lunch hour tomorrow, take an extra twenty percent off the sale price,"* said another. In fact, the average prosumer received just under forty targeted e-mails during the holiday season. The fact was that just under fifty percent of customers intended to do at least some of their shopping on-line in that year. The numbers increased significantly in each subsequent year.

Automation allowed every aspect of a potential customer's response to be scrutinized including where the e-mail was opened and exactly which words and/or pictures made a client from a particular demographic click on the *"buy"* button. The competition was fierce: Out of all of the targeting e-mails sent, just over twenty percent would be opened by the target and under ten percent of these would result in a "click" to get to the desired website and of these, less than two percent of these actually made a purchase. (Holmes 2012) The click-rate could be significantly increased by proper choice of an irresistible subject line. Vendors had the choice of sending huge numbers of older style "market-saturation" messages, or they could send smaller numbers of carefully tailored, custom generated, personalized messages that were entirely based on long-term studies of a particular individual's previous browsing and purchasing behaviours.

To survive, advertising had to integrate itself into an all encompassing grid that linked together web-based social media, smart-television, interactive-billboards, magazines, e-newspapers and e-zines. What exactly was an interactive-billboard? A dumb billboard just sat there with one fixed picture delivering a constant message to everyone who cared to look. The problem was that increasingly, people were tuning out and were *NOT* looking at dumb billboards anymore. A smart billboard was equipped with a high resolution LED screen, a video camera, internal computer and a high-speed wireless internet link. When a car approached the billboard from a distance, the camera focused in on the car's license plate and on the face of the driver. Accessing the cloud-based data base, in less than a millisecond the billboard identified the likely identity of the driver and using the knowledge of the individual's age, sex, demographic, education and known previous product and service purchases, the billboard instantly altered its projected image to one that is more likely to attract the interest of the passing motorist.

The real impact of Just-On-Order-Media, however, came from a technology called Augmented Reality (AR). Essentially AR used various display technologies to add custom information to the perceptions produced by the normal senses. Most people knew something about Virtual Reality (VR) that attempted to create completely artificial worlds that people could immerse themselves into. AR was different in that, with permission, it enhanced and was fully integrated with the information being supplied by the senses.

You were going for a stroll down a busy commercial street in a typical city. Even though your vision was 20/20, you put on a special pair of glasses that were linked to the wireless controller in your pocket. While still allowing light from the real world to reach your eyes, the glasses had prisms that reflected computer generated graphics onto your view of the street in front of you. The glasses were equipped with a complex series of sensors that kept track of your exact geographical position by GPS and using optical gyroscopes, they could keep track of the orientation of your head. In this way the AR graphics could stay properly positioned and superimposed onto the real-world scene. (Feiner 2002)

At the beginning of the stroll, you turned on the unit by pressing "activate" on the controller. From that moment forward, as you walked past various commercial enterprises along the street, real-time advertisements appeared as HUDs (head up displays) on the inside surface of the glasses. The interesting thing about these ads was that they were semi-transparent, so you could see through them to observe the real world of sidewalks and storefronts. As you walked by a shoe store, a pop-up ad informed you that there was a special sale on men's shoes that would end in ten minutes. As you walked past the next commercial establishment which happened to be a travel agency, a new pop-up ad told you about a special all inclusive holiday package in the Caribbean - but only if you walked in and booked it within the next five minutes. And so it went.

Now imagine that you were a tourist in a historical city such as Rome. The AR system could present you with 3D models of long since vanished buildings overlaid at their original locations. Virtual time travel became possible as you explored the forum as it looked at the time Augustus, Hadrian or during the time of the Borgia's.

Figure 4-1: A woman wearing glasses with the augmented reality HUD (head up display) activated. This could also be the look of cell phones of the future. Source: Leonard Low

As long as the system was enabled, it continued to feed you real time information about what was happening in the world around you. By superimposing a rich, constantly evolving virtual world of information onto the real world, your ability to stay current with unlimited streams of fresh knowledge become virtually limitless. The application of various filters to this information storm was of utmost importance to keep from being drowned by an information deluge.

There were already a number of AR applications available for smart phones with many more under development. Driven by the phone's GPS and compass systems, some applications saved users the trouble of weaselling out information about the ratings and reviews on restaurants in a given area while others drew on relevant information from selected on-line resources such as Expedia or Wikipedia. Other apps used an image taken by the phone's camera, of say, an alleged priceless antique that someone was trying to sell to you, and that app would automatically search all of the relevant data bases to deliver all the information on the item so that a decision could be made to purchase based on high-level, real-time intelligence, rather than on emotion and alleged information.

More advanced versions of AR systems projected virtual images of products in front of your eyes where you could actually interact with them. This meant that if you were examining a virtual jewellery box, you could actually reach out and turn it around to reveal the back, the bottom and then actually open the lid to look inside to see the contents of the box.

Beyond AR there lay the rapidly developing world of Immersive Reality (IR). IR was a highly interactive 3D technology that was able to create an illusion of depth and perspective by use of stereoscopic projectors illuminating images in a specially adapted booth or chamber. Professionals such as engineers, architects, fluid dynamics experts and realtors began using IR to create realistic virtual renderings of systems of interest so that they could fully investigate the space as if they were actually there.

An organization known as Reality Cave Inc. invested more than $5 million to open a 3D design centre in the region of Kitchener, Ontario to meet the high demand for accurate visualization of complex systems during development. The purpose was to identify design errors so that they could be corrected before construction began. In the case of buildings, the architectural plans could be scanned and processed by the specially written software that converted the plans into a virtual 3D rendering of the entire building so that the user could interact with it in any way that they chose. This technology also opened up a whole new venue for advertisers who saw the opportunity to show the smallest details of products such as mobile homes or recreational properties to potential purchasers before they were built. A 3D - IR presentation of an apartment could be so realistic that users could clearly see vivid colours and room dimensions, while also "feeling" the textures of the floors and walls. Municipalities found that could use the Reality Cave to allow proponents to visually show citizens how neighbourhoods would be impacted by proposed real estate developments.

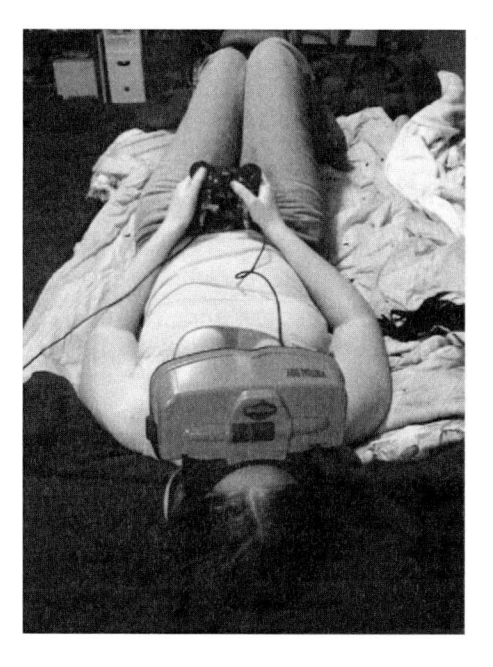

Figure 4-2: A woman playing an immersive video game using a first generation virtual reality head set. Virtual reality technologies are developing quickly and will play an increasingly important role in our lives in the future. Source: Kevin Simpson

Clearly, technology was trending on many fronts towards the provision of real-time information to anyone who needed it, via something as small as a pair of contact lenses. It was foreseen that there would soon be numerous applications of this technology for law enforcement officers, fire fighters, maintenance engineers and professional athletes. (Rockel 2012)

Of course, the technology long ago progressed beyond the point where serious privacy concerns had become obvious. Applications would soon become available that would allow anyone to point their phone at virtually anyone and to instantly be able to find out everything about them such as all the listings in their public records and social media accounts. It would also be possible to access private information that wasn't generally available.

What happens when privacy no longer exists?

—

Joy's Law, namely, that no matter how many smart people work for you, the most intelligent people work for someone else, applied just as significantly to the advertising industry as it did to other sectors. Traditionally, the professionals who worked for the world's biggest brands created ads with absolutely no input from the intended audience. However, in order to prosper in the 21st century, the public was gradually solicited to give greater and greater input into the creation of advertising campaigns.

Doritos launched a campaign in 2012 called "The End" that invited the public to write a suitable conclusion to an ad that they had created. The eventual winner was awarded $25,000 and the Doritos marketing team was left with a new respect for the latent creativity of their product consuming public. What was discovered that while consumers were often suspicious of ads created by industry professionals, they rapidly embraced Just-On-Order-Marketing-By-Peers as being much more trustworthy. After all, who was a person going to believe: the movie studio that was trumpeting a newly released movie as being *the best movie of the year,*" or her next door neighbour who had seen the movie and called it a "dud?" One thing that contributions from the public did really well was to help build excitement for a product through social media such as LinkedIn, Facebook and Twitter. In fact, the on-line attention paid to a brand evolved to become the most important part of the campaign. By letting the public take increasingly important roles in advertising campaigns, brands were successfully making a strong connection with their audiences and becoming an important part of the lives of consumers. (Krashinsky 2012)

Just On Order Movies

For almost a century, the business model of the film industry was based on keeping filmmakers isolated from the audiences they served. Once a film was completed, the studios and distribution companies controlled everything from the distribution and marketing of the film, to where and when it would be cleared for showing in the theatres. The rise of social media and Internet-based streaming services fundamentally changed everything, except perhaps the minds of the senior executives of the organizations that had benefited from the old system. For the first time the Internet empowered filmmakers to not only create, but also to market, distribute and show their films.

Just as the studio system had a strangle-hold over the production of movies, the theatre operators had absolute control over the distribution and presentation of the movies to audiences. The large and powerful theatre operators were able to keep the small independent distributors out, preventing many quality films from ever being seen by major audiences. (Hansen 2011)

Gradually, advances in video streaming technologies opened up the possibility that home audiences could watch movies on the same day that they opened in theatres. Content with the highly profitable status quo, the large theatres retaliated by boycotting all the movies that were to be released through Web-based, on-demand platforms. Certainly, online movies could not compete with the quality of TV or big-screen cinema movies, but all information, especially media information, desperately wanted to be free. The pressure was gradually building to have movie titles released to the public in the same week as they were first released in the theatres, creating a comprehensive system of Just-On-Order-Movies-and-Media.

As technology changed the way movies could be viewed by the public, it was also dramatically changing the way movies were made. The Oxford based media company, NaturalMotion, developed a program that used neural networks and synthetic evolution to generate self-animating, virtual entities capable of realistically walking, climbing, running and interacting with each other. As these virtual actors (VAs) became increasingly realistic by being programmed to move and react like real human beings when they were immersed into any particular environment, they began to threaten the jobs of stuntmen and other specialized actors in the movie industry.

Biologists noted that real-life evolution combined the emergence of new traits with adaptation to changing environments. In complex systems such as the evolving of a species in the wild, unexpected properties could and did often arise spontaneously and nature subsequently subjected these properties to powerful natural selection forces. (Morton 2004)

Genetic algorithms and neural networks were being used to simulate this complex process of natural selection, allowing virtual entities to essentially "evolve" with each subsequent generation. But instead of requiring decades for each generation as real living beings would, the computer based virtual life-forms could live out twenty generations or more in about five minutes. The developers of this technique found a way to make animated characters move like real ones, without any of the tedious manual effort that was required to achieve this result in the past, using a technology such as stop motion photography. Constant improvements in the software eventually allowed the exact reproduction of virtually everything, including the tiny identifying characteristics that made the motion of each particular actor unique. The goal was to create movies and interactive games that were populated with virtual actors that were essentially indistinguishable from real ones and that, most importantly, did not belong to the actor's union.

The granting of increasingly realistic characteristics to virtual actors was greatly advanced by the rapid evolution of digital scanning technologies. A number of famous actors already had all of their relevant characteristics digitized by a laser scanner with an accuracy exceeding a tenth of a millimetre. It was disturbing to some audience members to see their favourite human actor reduced to a few hundred million points in a 3D data base. But it was even more disturbing to then have these data points manipulated by a super computer to produce overall action so realistic that on the screen the virtual actor was virtually indistinguishable from the real one. In this world, actors were no longer compensated for their ability to deliver emotion-charged lines combined with powerful body language. Instead, the human actor received a royalty each time their digital facsimile appeared on the screen. (Gravity Design Studios 2014)

These virtual actors never got old, sick or temperamental. (Are you listening, Marilyn?) They could be made to carry out the most complex and dangerous stunts in complete safety. The complexity of the actions that the virtual actors could be made to do was only limited by the imaginations of the script writers. These Just-On-Order-Actors enabled the creation of Just-On-Order-Movies that would turn the industry on its head. As this technology trickled down to the home PC user it would enable increasing numbers of "ordinary" people to make feature length movies, with some of them winning Oscars for their efforts. But when a virtual actor won an Oscar for best "actor" who would pick up the award? The writer of the screen play? The programmer? The actor who was originally digitized and had his 3D data points used by the computer to animate the virtual actor? This new world was full of such intractable questions. Such is your JOOM destiny…

Only a relatively short time ago, the modeling of life-like 3D figures required expensive scanning lasers and image capture software. The Do-It-Yourself (DIY) movement was able to move into the creation of virtual actors and 3D figures with the help of software packages of increasing sophistication such as Autodesk's Photofly. Much of this software was made available for free to anyone who wanted to take the trouble to download it. What made this program unique was that anyone who had a digital camera could use it. The program was designed to use photogrammetry, meaning that it was able to take a series of 2D digital pictures showing an object from every direction (about one picture every five degrees), and then using the computational power of the Cloud, it was able to stitch all of the supplied pictures together to create a seamless 3D object. The early versions of the software tended to be crude because the system could be easily confused by the presence of shadows, poor lighting, uneven backgrounds or reflections. The result was a 3D model that had gaping holes or dramatic irregularities. The early software was certainly not accurate enough to create replacement spare parts, but it could be used to place a loved-one's head onto a favourite 3D action figure in a video game.

Most people who went to see feature films were only vaguely aware of the fact that these projects cost hundreds of millions of dollars to produce. But technology marched relentlessly on and by the second decade of the 21st century, the majority of households had acquired the basic equipment necessary to produce a video of passable quality. A film enthusiast named Ingrid Veninger put up $5,000 of her own money to establish the first $1,000 Feature Film Challenge. In addition to the prize money, the winner was offered free post-production of the work and a secured screening at The Royal Cinema in Toronto. In the words of Veninger:

> *"You don't have to wait for permission from anyone to do your art. You can put $1,000 on your credit card. You can make a feature that can play at festivals and hit the big screen. And if we can keep costs low enough, we can make money on these films."* (Everett-Green 2012)

Creative individuals were becoming increasingly comfortable pitching their ideas on crowd sourcing sites such Kickstarter. If the "crowd" thought that their ideas were good enough, they got the funding they requested along with a free market evaluation. However, experience had shown that more money did not necessary result in a better product. If the "1K Wave" was any example of what could be done, then the film industry was certainly quaking in its boots as the big screens were increasingly being taken over by the works of talented amateurs.

Stop-motion animation was a staple technology of the movie industry since the earliest beginnings of film in the 1920s. The introduction of

3D printers early in the 21st century brought a welcome breath of fresh innovation into this 100+ year old industry. The stop-motion movie *ParaNorman* was the first to use 3D printers to create all of the puppets faces and body parts. Another movie called *Caraline*, used tens of thousands of 3D printed parts. The face of each puppet was constructed of more than forty - 3D printed parts. By making each of these facial components interchangeable, it was possible for the creators to portray literally millions of different facial expressions on each of the puppets. (Roper 2012)

The Video Game Industry

While the figure was not known exactly, by the opening decade of the 21st century it was estimated that the value of the global video game industry exceeded $65 billion US. At about the same time, it was rapidly becoming apparent that the dominance of the stand-alone gaming system was rapidly coming to a close, to be replaced by simpler, Just-On-Order-Games that were available from The Cloud, and that could be played on a variety of hand-held devices such as iPhones, Android Phones, iPads or tablet computers. Nintendo attempted to counter this trend by introducing to the market much more powerful gaming systems such as the 3DS, but sales of these devices remained far less than anticipated. This was a strong indication that the era of Just-On-Order-Games had arrived, and that this new business model would likely become the dominant one for the foreseeable future. (El Akkad 2012)

In contrast to the difficulties suffered by Nintendo, Microsoft's Xbox 360 system experienced continuing success due to the continuous release of new accessories, but especially because of the availability of cloud-based features that could continuously expand the capabilities of the console. This meant that the Microsoft system, along with Sony's successful Playstation console, could really be considered to be Just-On-Order-Media-Hubs that could also be used to download movies and other dynamic content instead of being simply a static gaming system.

These JOOM Hubs essentially become gaming alternatives to smart TV systems.

An idea that began to gain traction in the second decade of the 21st century was the concept of the "Render Farm" consisting of thousands of dedicated servers which offered Just-On-Order-Computational-Power to companies wishing to launch large digital media projects. The companies simply moved in and rented exactly the computational assets that they needed when they needed it. When the project was completed, they could simply walk away and leave the space to another start-up. Another great advantage was that participants who experienced a temporary reduction in their server needs could sub-let any excess server capacity to other organizations.

As a one hundred percent internet based company, Zynga was unique in that it started operations with a leased data centre space and continued to use leased space for up to eighty percent of its operations. Its goal was to find the most efficient infrastructure for its multi-million user on-line game hosting operations. Zynga was the author of more than half a dozen successful on-line games including Hidden Chronicles, Zynga Poker, Mafia Wars, Castle Ville and Farm Ville. Zynga was the first to configure particular servers for specific functions such as database or Web hosting. This system, which the company called its zcloud, was able to handle more than three times as many active users as other server systems. (Babcock 2012)

In the past, start-up companies were severely constrained by the requirement to raise all of the funds necessary to purchase data centre capacity in advance, just in case their flagship product became a best seller. With the advent of Just-On-Time-Cloud-Servers, if the number of people using an on-line game began to grow exponentially, the server capacity could be expanded at the same rate, virtually indefinitely. In the case of Zynga's on-line game CityVille, it reached more than 20 million users in less than five weeks. By the fall of 2012, all of the company's offerings entertained almost 300 million users per month. The managers of the company knew that to keep their customers coming back, they had to keep offering new, better, more creative and highly original games. The programmers were able to maintain their high levels of creative output by *not* re-inventing the wheel for every game. In new games, as much as possible, new features were often based on the tested game logic used in an older hit game. Furthermore, the zCloud environment offered a tool called zOps which allowed developers to see how well the new games were doing as they worked in the actual production environment.

Another important application of the technologies originally developed for the video game industry was in the creation of Virtual Worlds for various entertainment and medical applications. (Tinari 2009) Advances in graphics animation and projection technologies permitted the creation of fully immersive environments that incorporated most of the elements found in the real world such as furniture, plants and accessories such as kitchen utensils. As an example of a useful medical application, it was found that placing burn patients into an immersive environment filled with snow and ice significantly reduced the intensity of their suffering. Allowing hospitalized children to immerse themselves into a virtual environment that looked like their bedroom at home significantly sped up their healing process. Medical equipment could be made to appear like friendly toys so that children could become familiar with it before medical tests and procedures were performed. When a syringe was made to appear like a toy rocket, it was far less threatening to young patients. Transforming medical staff into non-threatening avatars was also found to be a significant way of reducing the stress levels of young children forced to undergo extended medical procedures. Avatars could also interact with children during their post surgical rehabilitation, by asking them to mimic various exercises. Trials have shown that many children would be more willing to follow the examples of such virtual adults than of the more threatening medical personnel that would normally be at their bedside. (Viatteau 2011)

It was of great concern to the US armed forces that in the first six months of 2012, more active duty military personnel committed suicide than were killed in combat. Contributing to this troubling statistic was the fact that soldiers experiencing difficulties were reluctant to seek help because of the not completely unrealistic perception that this could adversely affect their careers. The solution that was found to be highly effective was to create a virtual world populated by Just-On-Order-Mental-Health-Guide-Avatars that could be used to anonymously deliver psychological advice to anyone requesting it.

SimCoach was a second generation artificial intelligence (AI) program that represented a quantum jump in performance over previous programs. It was able to maintain records of all of the answers provided by a particular individual and to subsequently formulate increasingly complex customized responses. In other words, the program could evolve along with the soldiers that it was helping. Another program known as SimSensei was designed to monitor vocal stress patterns, facial expressions, body posture and hand gestures to help the virtual young woman avatar in a green aviator suit to formulate appropriate responses. Previously, there were long waiting lists of veterans awaiting counselling, but with the installation of the program in widely deployed kiosks, hundreds of soldiers were able to experience Just-On-Order-Mental-Health. The next improvement was the installation of the program into The Cloud so that individuals could access the service through their home Internet connection. (Piore 2012)

The highly seductive video game known as Skylanders racked up more than $200 million in sales in North America in 2012 alone. What made it so lucrative for its creators was the fact that it not only consisted of a videogame series, but also a line of cute and highly collectible toy figurines. To the despair of parents, their children needed to keep buying additional figurines to access higher levels of the game, and when they got to the store, they were likely to have found that all of the most desirable figurines were "sold out."

The company had found that by keeping the supply of figurines tightly controlled, it was able to keep customer demand high. However, the artificial restriction of supply was not able to withstand the increase in availability of low cost 3D scanning and printing equipment that allowed anyone with a complete set of all forty eight characters to make a detailed copy of each one, post the 3D designs on a site such as **iHive3D.com**, and then allow everyone on the planet to download and replicate them. Yes, the company could launch a legal challenge, but suits could only work their way through the legal system so fast, and in the meantime, millions of users would be downloading illegally made copies of the figurines. For manufacturers, it was a whole new world fraught with dangers and lessons of what could happen if they attempted to unnaturally manipulate their customers buying behaviours.

The Future of Print Media

The advent of the automobile did not completely eliminate the horse, it just turned it into a niche product. Similarly, technology transformed vinyl records and printed newspapers into niche products, but did not completely eliminate them. The rise of radio in the early 20th century forced newspapers to speed up their delivery of the news while the arrival of television later in the century forced them to incorporate many more pictures into their daily products. The rise of the Internet ushered the era of 24/7 news, and newspapers had to adapt once again.

One of the first indications that business in the book industry was changing was when Indigo Books & Music Inc. announced that it was going to launch an in-house private label publishing service, in effect becoming a competitor to many of its own supplier-publishers. The company saw the introduction of self-publishing as a move that would allow it to offer lower prices on low-volume, speciality Canadian books. At the same time, a number of book sellers dramatically moved away from having to stock huge inventories of books, traditionally kept in inventory just in case there could be a demand for them. They accomplished this by introducing on-demand publishing systems such as the Espresso Book Machine into the store that was capable of printing a "one-off" printed copy of a book when a customer ordered it. This Just-On-Order-Publishing technology was able to offer the consumer dramatically lower prices on printed books, while at the same time reducing costs for the book seller, who did not have to use valuable space stocking inventories of books.

The development of rigid plastic/metal e-readers such as the Kindle certainly had a great impact on the way in which written material could be consumed. However, many people maintained a preference for reading the news on a sheet of paper. This was why the development of the first interactive "paper" computer by the Queen's University Human Media Lab marked the true beginning of the end of the paper age. What made this computer unique was that it looked, felt and performed like a sheet of 8.5" x 11" paper, but it was actually a multi-use computer that could not only be folded, rolled up and written on, but also would eventually be capable of storing over 100,000 book pages while still doing everything a laptop computer could do. (Reviewgoods 2010)

Many years ago, the author first outlined his observations on similarities he had found in the patterns associated with the evolution of new technologies in different domains. (Tinari 1978) In the years since, these observations were distilled into what some have termed Tinari's Law of Paradigm Shifting Technologies. In simplified form, it basically states that: "All fundamentally new technologies initially do the opposite of their ultimate, long-term trend." One reason that this was true was that new technologies needed some time to "mature" before they could work as originally intended. A second reason was that users tended to try to force a new technology to work within the old technological paradigm, undermining any potential advantages it could offer. An example of this Law in action was with the introduction of the first cannons to the battlefield in the 1400's. While they were scary looking, they were so poorly constructed that they were actually more dangerous to their own firing crews than they were to the enemy. In another example, the first automobiles were so unreliable and failure prone that they were actually a worse transportation option than the technology they would ultimately replace (the horse). In a more recent example, computers were meant to usher in the golden age of the paperless office. In actual fact, computers combined with high capacity laser/ink jet printers led to an explosion in paper use as office workers used them as little more than high speed typewriters. However, many years after the first introduction of computers to the office, the technology finally matured to the point where it could finally manifest its proper long-term beneficial trend.

The widespread adoption of these foldable/rollable, interactive paper-like computers would finally put an end to tree-sourced paper and allow the creation of the truly paperless office. From the point of view of those in the industry, this was a disaster in slow motion because this technology would mostly put an end to the market for laser/inkjet printers and would greatly upset those holding stock in industries such as pulp and paper. Since the new technology could also be used to construct new generations of foldable, paper thin communication devices, it could also put an end to the smart phone industry as it originally conceived. This is yet another aspect of your JOOM destiny...

—

In 2009 there was a meeting in Montreal of leading newspaper executives to discuss the future of the industry. It was agreed that people would always be interested in finding out about local, national and international news, but the question was in what form people would be receiving it. The first generations of e-readers were incapable of displaying colour pictures, multimedia or high-speed, on-line content. These limitations were gradually eliminated in subsequent reading devices, allowing for the full integration of the written word, sound, pictures and video.

The current revolution will not only completely change the nature of newspapers (what they are made of) but also their content. In the era of mass communication, one newspaper was printed for all subscribers, and they had to take it or leave it. With the rise of digital content, national newspapers began printing different papers for each region such as having content tailored for the west coast, the interior and the east coast. The next stage was the introduction of Just-On-Order-Media where each subscriber specified their main areas of interests and whenever desired, a unique, custom designed newspaper would be assembled for each individual subscriber.

At one time, putting together each issue of a major national daily newspaper required a staff of hundreds of highly specialized individuals, including reporters, writers, editors and printing-press operators. In contrast, the JOOM Newspaper would be assembled automatically from information stored in on-line data bases and delivered to the subscriber without any outside human intervention. Clearly, the impacts on employment in this industry alone will be highly significant.

Right from their first appearance at the beginning of the 21st century, newspaper websites attracted an exponentially growing viewing audience. By 2009 these websites received almost seventy million visitors, consisting of more than forty percent of all users of the Internet. But exactly what business model(s) could the newspapers adopt that would permit their professional journalists to work successfully with thousands of on-line citizen journalists, publishing their work alongside the virtually limitless information available from social media while remaining relevant to the majority of their subscribers as technologies continued their rapid evolution? (Joel 2009)

Just-On-Order-Media for Education

According to the well-known expert in education, Sir Ken Robinson, the most important thing that a school should do was to teach students the most effective way to learn. According to Robinson:

> *"Personalized learning is the process of contouring learning to the individuals that you're dealing with, recognizing that we all have different strengths and weaknesses, different interests and different ways of learning...It isn't that everyone has to learn different things , although eventually our interests will take us in different directions...But in terms of the things we want all people to learn...personalized learning is finding the best ways to engage with people with different interests, passions and ways of thinking."* (Steffenhagen 2011)

Robinson produced a popular TED talk, "Do Schools Kill Creativity?" that was watched by more than 200 million people from around the globe. The essence of his message was that all education should de designed to recognize the whole person. He claimed to have recognized that most attempts by governments to reform education failed because the changes were implemented in a command-and-control fashion where teachers were given exact instructions on what they were to teach and how they were to teach it. The direct result of this according to Robinson was the narrowing of curriculum, a focus on those things that could be easily measured and a mass impersonal education rather than custom tailored, personal education. When asked if he had any suggestions for teachers to improve their craft so that they could deliver a personalized learning experience to each student, Robinson suggested:

- Teachers had more freedom to enhance the curriculum than they thought

- Teachers should be doing the job because they loved it and because they saw their work not as just a job, but as an art form

- Teachers had to be prepared to learn from students

As an internationally known consultant in organizational creativity, the author proposed to the provincial Minister of Education in British Columbia, Canada to develop a series of boxed "creativity modules" that could be used by students in K – 12 classrooms to enhance the levels of creative thinking within the traditional school system. The proposal was accepted, and it was expected that students in a number of jurisdictions would eventually have the opportunity to be exposed to hundreds of mind expanding, Just-On-Order-Modules designed to enhance innovative capabilities and creative thinking skills.

In 2011 the British Columbia Ministry of Education announced that each student would be encouraged to bring their laptops, digital tablets, smart phones and other electronic media to their schools. This media would be used to deliver to each student Just-On-Order personalized learning plans for every student. While the intention of the program was obviously good, no additional money was specified by the Ministry to implement the program, so each teacher would have to carry it out to the best of their ability within the existing budget. According to the Minister of Education George Abbot:

> "Many of the opportunities and jobs we're preparing our students for don't even exist today...So while we enjoy a strong and stable system, we need a more nimble and flexible one that can adapt more quickly to better meet the needs of 21st Century learners...A curriculum with fewer but higher level outcomes will create time to allow deeper learning and understanding...Increased flexibility will be key to making sure the student's passions and interests are realized, as well as their different and individual ways of learning." (Hunter 2011)

Just-On-Order-Journalism

After the recession of 2008, hundreds of top quality journalists were downsized from their main-stream media jobs and were forced to become independent agents. This resulted in the invention of a new media profession that became known as Brand Journalism. Brand Journalism was meant to describe the creation of a compelling story about what an organization could do, while at the same time convincing people that they were not being "sold." These individuals could be found writing content for the hundreds of free magazines, such as those that an air-sick traveller might encounter while desperately fumbling for a barf-bag in the back-seat pocket of an airline seat.

Even more importantly, it was a stark fact that consumers were increasingly making significant buying decisions by searching on-line. Increasingly sophisticated Google search protocols tended to select sites that had the newest and most relevant content. This meant that simply to get noticed, each company had to have professionals on staff who knew how to write compelling content that would keep the organization's products at the top of Google's search pages. The powers that be within the corporate world realized that their customers were increasingly to be found on-line and if they were to be drawn to the company's wares, there had to be constantly renewed, fresh, helpful and interesting information there for them to read. The concern was that it was becoming increasingly difficult to differentiate between the pure objective news stories and the sponsored content that talented journalists were paid to slant into a specified direction. (Basen 2012)

Throughout the 20th century, the huge costs associated with professional cameras and video equipment meant that most news stories could only be properly covered by professional journalists with their video crews. Even when amateurs directly witnessed a developing news story, the difficulties and slowness of communicating information meant they could rarely get their work published in the mainline media.

In the 21st century, the almost universal availability of camera-equipped smart phones with Wi-Fi Internet connections linking to Internet-based instant news and video posting sites meant that for the first time every citizen had the potential to report on news stories with the same potential impact as the pros. At the same time, the capacity to load and store video records of events in sites such as YouTube meant that events that in the past would have been reported once and then gradually forgotten, were now being placed into the public record forever.

Nearly everyone remembered that President Bush continued reading to elementary school students on September 11, 2001 after he had been told of the first impact of an American civilian airliner into the twin towers. The bizarre image was shown again and again as citizens attempted to make sense of the national tragedy and to decide if their commander in chief was up to the task of dealing with it. Modern politicians were under the strict control of their handlers and *each* public utterance, appearance and deed was carefully scripted to convey a specific message, while at the same time placing the candidate in the best possible light. From the point of view of the handlers, the most stressful times occurred when their charge had to leave the structured scripts behind and sit down one-on-one with a professional journalist and to actually generate coherent answers to relevant, off-the-cuff questions.

At one time, if the politician's performance in such an interview was less than stellar, the negative impacts would have been minimal because only the spectators watching the actual "real-time" interview would have been aware of the gaff. In the new world of Just-On-Order-Media however, the interview would immediately be posted on YouTube for potentially the entire voting population of the nation to download and form their own opinions.

Many believed that the era of universal media began with the amateur recording of the Rodney King incident in Los Angeles in 1991. After this incident, there followed a long series of recordings of the abuse of private citizens by agents of their government such as the police. As only one example, anyone could relive the abuse of a woman who was given a humiliating body cavity search during a routine roadside stop. The URL of the incident could be seen at: http://www.youtube.com/watch?v=oONF84y1cx8

Just-On-Order-Art

At the entrance to the Museum of African American History in the Smithsonian in Washington, there was a bronze statue of Thomas Jefferson. Few of the visitors knew that this statue was actually an exact copy of the original in Monticello. The way the copy was made represented a fundamental paradigm shift that would eventually impact every corner of the art world and shake it to its very foundations.

To make a copy of a priceless statue such as Michelangelo's David in Florence, all that was required was a portable laser scanner, a computer loaded with 3D editing software and a 3D printer capable of working in the desired material. The 3D scanner was used to digitize the statue and the resulting file was transferred to a 3D editing program, where it was converted into printable form by slicing it into a multitude of thin layers and then the edited file was downloaded to a 3D printer for production.

Until recently, there was a significant barrier to entry because the required equipment cost hundreds of thousands or even millions of dollars. With time however, driven by its own version of Moore's Law, prices for scanners, supporting hardware and 3D printers fell by about fifty percent every eighteen months. In fact, some perfectly functional software such as *123D Design* by Autodesk was made available to anyone as a free download. This meant that the technology that was once unattainable except for 3D design professionals, soon became available to anyone with an interest in using it.

As another example of increased accessibility was a free program released by Autodesk called *123D Catch* that allowed non-technical users to create a detailed 3D scan of an object simply by taking pictures of it from several different directions and angles. It is only a matter of time before all of the greatest works of sculpture now stored out of sight in museum basements were scanned and then, forever thereafter, they could be made available to anyone with a tablet computer or a smart phone.

For painting, an interesting question centered on exactly what were the intangible characteristics that made a particular work a masterpiece. Anyone who walked the halls of the Louvre knew that it was much more than the image represented by the painting itself, but it was also the rich texture of the paint on the canvas and its three dimensional undulations that represented the raw emotion of the artist, fired in the forge of artistic creation. Previously, the great works of art of the world were isolated away from the public up on pedestals or behind protective glass. Now with 3D printing, it was possible to create exact copies of Van Gogh's angry and substantial brush strokes that students could actually run their fingers over. For the first time in history, the most insignificant small town gallery could have hundreds of the best pieces from the Vatican or any of the world's great collections and these works would each be such good copies, that an art historian would need an x-ray machine to differentiate the copy from the original. (Hopper 2012)

How big a threat will technology be to the traditional art world? According to literary critic Walter Benjamin, original works of art were:

".....imbued with a unique aura that withered when they were replicated and that by making many reproductions it substituted a plurality of copies for a unique existence." (Benjamin 2012)

No matter how many copies of the pages of Leonardo's notebooks a historian reviewed, there was still nothing like the total bodily and emotional excitement that occurred when she put on the pair of white gloves, removed an original sheet of paper from its humidity controlled storage case, and then actually handled a sheet that 500 years ago the master had actually touched with his pen and elaborated with his now famous "mirror" handwriting.

Emerging technologies opened up many opportunities for visionary artists. Toronto-based Evan Penny used a 3D scanner to create detailed images of human subjects. After all the data was inputted, the artist carefully "skewed" the image just enough to create a "creative distortion." Experiments showed that if the original image was too heavily distorted, it rapidly became unrecognizable as human. When the artist was happy with the slightly modified 3D image of the subject, the data file was sent to a Computer Numerical Controlled (CNC) Milling Machine which produced the hard foam core of the statue. The artist applied a layer of modelling clay that was then sculpted into the desired image. Finally the image was moulded and cast in silicone. Penny's work was exhibited in a number of museums around the world including Museum der Moderne in Salzburg, Austria. What was certain was that 3D making and the Just-On-Order-Art that it could produce would soon have profound impacts on many sectors of the world of modern art.

Just-On-Order-Maps and the Internet as a Media Enabling Technology

The University of Southern California's Visualizing Science Initiative was founded with the mandate to find ways to better illustrate scientific concepts by the use of visual media. In one project, a 3D printer was used to construct a scale model of the Yosemite Valley and then thousands of overlapping pictures were projected onto the digital terrain to create a truly realistic representation of that beautiful California wilderness area. Digital models were also created of Catalina Island and of the surrounding sea floor for biological research. Researchers also attempted to find new ways to model and illustrate climate change in 3D.

The Internet was the greatest facilitator of media dissemination ever conceived. Sites such Google, iStockphoto, YouTube and iHive3D. com have made maps, pictures, videos and 3D art works available that once would have sat for years in dusty archives. Just-On-Order for anyone with a wireless connection. The rapid proliferation of miniaturized computational capabilities, motes (tiny remote semi-intelligent sensors), IP (Internet Protocol) capable devices, Radio Frequency Identification (RFID) chips, micro-sensors, ever cheaper wireless networks, automated machine-to-machine (M2M) communications, apps for smart phones and cloud computing led to what has been termed the "Internet-Of-Things" (IOT) or the Just-On-Order-Net. This was such an important development that it will be discussed in some detail in the following section.

Future Trends in Media

Right from the time of their first deployments, the problem with traditional computers was that they were just dumb boxes that could only do what they had been programmed to do. The way that computers were originally designed (i.e. von Neumann architecture) was fine for data processing and spread sheets, but did not work very well for speech recognition or creative problem solving. In order to develop machines that could adapt, learn and grow the way humans did, it was necessary to completely re-invent the mechanisms by which computers operated.

The Canadian company, D-Wave Systems, developed a computer whose software was not fixed in memory, but that was able to re-write itself to incorporate the results of its day-to-day experiences. As it interacted with humans, good behaviours were reinforced while bad behaviours were discouraged, thereby promoting learning. The D-Wave machines disposed of the traditional transistors and integrated circuits. Rather their basic building block was the Q-bit that did not operate according to the laws of classical electronics engineering, but according to quantum mechanics. The processes used in these machines were so powerful and efficient that a one hundred and twenty-eight Q-bit chip already had a computational capacity that exceeded the supercomputers of the era. The plan was to continuously expand the power of these devices until they exceeded one million Q-bits. What these computers will be able to do can only be speculated on but it is almost certain that they will be able to generate Just-On-Order research, creative writing, art, poetry, journalism, news stories, movie screen plays, language translation, medical diagnoses, academic and practical instruction and medical diagnosis.

The humans who worked in these various media professions will not have much to worry about in the short term. However, over the long-term, machines based on new architectures will become increasingly comfortable doing creative media jobs that humans once thought were safe from automated competition.

The Internet will be *the* defining and enabling "media" technology of the 21st century. The exponentially increasing number of devices connecting to the internet meant that the Internet Protocol Version Four (IPv4) was rapidly running out of address spaces. The solution was to move to the new platform known as IPv6, which offered more efficient routing, packet switching, directed data flows, simplified network connections, support for many new services and much better security. Most importantly, it offered 128-bit addressing meaning that it could support 2^{128} or more than 3 x 10^{38} addresses, or almost 10^{29} times more addresses than the previous IPv4.

This avalanche of addresses was soon to be needed to enable the Internet of Things (IOT) which will be first, the assignment of an internet address to objects, and second, enabling devices to easily communicate with each other. The IOT will initially consist of billions and eventually trillions of IP-enabled devices including tablet computers, smart phone apps, wireless sensors including internal medical body sensors and external networks consisting of motes (remote sensors), product RFID (radio frequency identification) tags, smart TVs, video games, mp3 players, home appliances, security systems, electrical smart meters, power outlets, vending machines, smart vehicles, child locator tags, micro-electronic mechanical systems (MEMS) and much more. The IOT will allow the monitoring of virtually anything including body metabolism, perimeter security, building heating, ventilation and air conditioning (HVAC) wireless LED lighting systems, industrial control and automation systems and infrastructure (bridges, rail lines, traffic lights etc.) monitoring. In other words, everything that can be monitored, eventually will be.

As an example of the IOT in action, on a summer day, once the external temperature of a room exceeded 33°C, one of the power outlets activated, turning on a fan and dispensing some water into a bowl on the floor, helping to keep the cat cool. The fridge, doing exactly what it had been programmed to do, e-mailed an order to the local grocery store for additional ice cubes, ice cream and popsicles for the kids. Sensors implanted in the various garden plants signalled that they needed attention and in response, the lawn sprinkler activated automatically, until the plants communicated that they had had enough. A message was sent from the owner's smart phone automatically purchasing stock in companies making summer equipment such as fans, air conditioners and sun hats. As prices for sensors and wireless technologies plunged, more case studies even more extraordinary than this one would soon become possible.

The above case study notwithstanding, at the time of this writing, it was impossible to exactly know what the ultimate impacts on human beings and on society would result from the universal Internetization of everything.

Conclusion

The mass publication of paper versions of books, magazines and news-papers will gradually be transferred to e-readers of steadily increasingly sophistication, while those quaint people still wanting paper copies of specialty titles will increasingly get them from print-on-demand systems. Eventually tablet-based home systems will be capable of pro-ducing Just-On-Order-Paperbacks. This will mean that every person with an idea will be capable of putting it into print, so crowd sourcing will increasingly be relied on to differentiate good from bad books and literature.

TV, movies and video will continue to evolve from 2D, to virtual 3D, to real 3D. This latter technology will involve the projection of true holographic images into space allowing families of the future to sit in a circle around their "TV" as they were actually "immersed" into the thick of the action.

Art galleries, traditionally involved with the presentation and sale of visual art such as sculpture, will have to dramatically change their busi-ness models. When a piece of art was sold, the artist commonly split the sale price 50/50 with the gallery. Each piece of art generated one lump sum payment for the artist, and no more. In contrast, if a sculp-ture was digitized with a 3D scanner, it could be posted on a site such as **iHive3D.com** and could potentially be sold thousands or even millions of times, generating a long-term residual income for both the artist and the agent. The numbers are extremely compelling and all galleries will eventually have to embrace the new model if they are to remain competitive.

The Internet has grown from the network of the few (DARPA-NET) to the network of the many (www) to the Internet of Everything (IOT) and it will continue to evolve until it encompasses virtually everything that matters. It will be the enabling technology that will act as the stage on which the evolution of all of the media technologies will play out.

A statement was made by a media observer with the introduction of the first talking movies in the 1920s that strangely enough, still applies today: "*You ain't seen nothing yet, baby!*" Get ready to embrace your JOOM destiny...

Web Resources

Virtual 3D Actor: http://www.youtube.com/watch?v=nSEOjzeF32o

Virtual 3D Woman: http://www.youtube.com/watch?v=VC5e1KDyb24

Virtual Sword Fight: http://www.youtube.com/watch?v=ySf4f19CNyU

Real Time 3D Animation: http://www.youtube.com/watch?v=bZ78loEhAmg

3D Print Your Baby Before Birth:
http://www.policymic.com/articles/79435/
you-can-now-3d-print-your-baby-before-it-s-born

Photo Credits

Figure 4-1: http://commons.wikimedia.org/wiki/
File:Augmented_reality_-_heads_up_display_concept.jpg

Figure 4-2: http://commons.wikimedia.org/wiki/File:The_
only_comfortable_way_to_play_a_Virtual_Boy.jpg

References

Babcock, C. (2012) "Zynga Redefines Cloud, Customer Service", *Information Week*, Sept. 17, pp. 22-23

Basen, I. (2012) "Is That an Ad or a News Story – And Does It Matter Which?" *Globe and Mail*, August 7

Benjamin, W. (2012) "*The Work of Art in the Age of Mechanical Reproduction*," Prism Key Press

El Akkad, O. (2012) "Can Nintendo Evolve From a Pure-Play Game Hardware Company in Time," *Globe & Mail*, Jan. 26, p.1

Everett-Green, R. (2012) "Can a Great Film be made for $1000?" *Globe and Mail*, October 9

Feiner, S.K. (2002) "Augmented Reality," *Scientific American*, April, pp. 49 -55

Gravity Design Studios (2014) "ELISSA, Variation Three. An example of a first generation virtual actor, she is available in a variety of flavors with regard to makeup, eyes, lip colour, eyelash length, face shape and hair." http://gravitydesignstudios.com/elissa-3.htm

Hansen, D. (2011) "Filmmakers Take Charge of Their Movies from Start to End," *Vancouver Sun*, September 29

Holmes, E. (2012) "The Dark Art of Store E-Mails," Report on Business, *Globe and Mail*, December 19, P. B8

Hopper, T. (2012) "Is it the Real Scream?" *National Post*, July 7, p. A20

Houpt, S. (2011) "In Sunny Riviera, Storm Clouds Gather," *Globe and Mail*, June 24, p. B8

Hunter, J. (2011) "B.C. Touts New Personalized Learning Plan," *Globe and Mail*, October 29, p. A23.

Joel, M. (2009) "The Newspaper of the Future," *Vancouver Sun*, May 22, p. C2

Krashinsky, S. (2012) "Advertising's Newest Ploy? Get Consumers to Make Your Ads," Report on Business, *Globe and Mail*, Feb. 10, p. B6.

Morton, O. (2004) "Attack of the Stuntbots," *Wired*, January, pp. 154-159

Piore, A. (2012) "Big Idea: Help Stressed Vets with Sim Coaches," *Discover*, December, p. 10

Rockel, N. (2012) "Let Your Eyes Do the Walking," *Globe & Mail*, Feb. 11, p. A19

Roper, C. (2012) "The Boy with 8,000 Faces," *Wired*, August, p. 93

Reviewgoods (2010) "Future Notebook, Flexible Display," Nov. 21, http://reviewgoods.blogspot.ca/2010/11/future-notebook.html

Seguela, J. (2010) "Leading Creative People," *European Management Journal*, Volume 28, Issue 4, August, pp. 278-284.

Steffenhagen J. (2011) "Teach Students, Not Subjects, Popular International Speaker Says," *Vancouver Sun*, August 24, p. A5

Tinari, P.D. (1978) "Self-Sustaining Habitations, Energy Efficient Design, and Views on Catastrophe Theory," Physics Colloquium, Department of Physics, Queen's University, January 12.

Tinari, P.D. (2009) "The Future of Virtual Worlds," NASA Special Presentation, April 30, https://www.youtube.com/watch?v=acCc38UcYYA

Viatteau, M. (2011) "Montreal Doctors Design Virtual World to Help Traumatized, Hospitalized Kids," *Vancouver Sun, December 28*

Chapter 5
The Global JOOM Revolution Just-On-Order-Impacts

Introduction

Most of the articles and books that appeared in the last twenty years on the subject of 3D making did an excellent job at presenting the details of the technologies that were driving this global revolution. However, these works said relatively little about the profound long-term political, economic, sociological and philosophical changes that would be driven by the widespread adoption of these technologies.

Looking at history, the widespread adoption of the factory system in England during the late 18th and early 19th centuries, caused massive social dislocations and even a significant counter-revolution in the form of the Luddite movement. But the long-term effect of the industrial revolution was the dramatic decline in the costs of many of the goods that ordinary people needed in their daily lives. These were the goods that they used to make themselves, but that were now being produced by the cartload in the new urban factories and then sold in local village stores. Both the factories and the stores were controlled by the newly emerged and rapidly growing merchant classes.

The ironic consequence of the 3D printing revolution, will be the virtual destruction of the factory system so laboriously built and refined by giants such as Henry Ford, and the return of manufacturing back into the home, where it was more than two hundred years ago. In this section, some of the consequences of this will be examined in some detail and projections will be made of some future trends in the industry and their possible impacts on the lives of ordinary people.

The Global Impacts of JOOM

The great recession of 2008 decimated many sectors of the North American economy. In addition to gutting the U.S. housing sector, one of the greatest impacts was suffered by the manufacturers of home appliances. The traditional business model that had sustained appliance manufacturing for a hundred years, suddenly became a huge liability. Stuck with huge inventories of unsold machines for which there was no longer any market, the large manufacturers were forced to close plants, dramatically reduce costs, and shed thousands of jobs to consolidate operations.

If these same manufacturers had embraced Just-On-Order-Manufacturing when times were good, they would have benefited from a business model that would have allowed them to eliminate all inventory by building an appliance *only* after there was a paid order for it. Reliance on external suppliers would have been eliminated, since all of the required components would have been manufactured on-site, only as needed, for each appliance. With no inventory and no possibility of an interruption of the supply line caused by the failure of a critical partner, the large appliance manufacturers would have found themselves in a much more favorable financial position after the great recession had done its damage.

The aviation industry also suffered deeply with the burdens of huge inventories during the recession in question. For example, by 2011

Boeing had accumulated more than $16 billion in 787 Dreamliner inventory. Many of the planes sat for months outside the plant in Everett, Washington, acting as a huge drag on the company's profits. During good times, if Boeing and other companies had gradually shifted their operations from their traditional to the JOOM business model, there would have been no inventory of aircraft, no major financial losses resulting from excess production, and no requirement to stockpile countless spare parts and components waiting for new aircraft to put them in. (Ray 2011)

JOOM was, by definition, multi-dimensional and was impacting multiple domains. Only a few of these were examined in this book. But in actual fact, there were many others that would soon be impacting the world.

As an example, all areas of technology were being impacted by the rapidly evolving field of Just-On-Order-Materials. Historically, the fundamental importance of materials to human development could be seen from the fact that major periods of history were named after the material that was predominately in use at that particular stage of development (i.e. the Stone, Bronze and Iron Ages). Moore's Law, that for over thirty years described the improvements in the number of electronic components that could be incorporated into a semiconductor chip, was essentially describing progress in materials science.

The real game changer will be the development of new generations of 3D printers capable of manufacturing materials at the molecular level. This will permit the development of new types of nano-composites with properties superior to any previously available material. This meant that materials could be created with properties designed for specific applications. Materials could be specifically designed with high tolerances for radiation damage for use in nuclear reactors, amazing flexibility for applications in bendable electronics, ultra-low coefficients of friction so pathogens could not adhere to surfaces for improved infection control and easier sterilization in health care applications, materials that self-regulated their mechanical, chemical or electrical properties in response to external stimuli such as temperature or pressure, and the creation of super crack-resistant materials for protective armour applications. (PS 2012)

Carbon fibre promised to be one of the materials that would characterize the 21st century. Carbon fibre was first manufactured in Japan and then shipped for further processing in the Western U.S. There, the basic fibre was transformed by a complex process into an incredibly light, but strong material with an almost limitless range of applications. The final step was to weave the fibre into a textile material that could be used, for example, to create an automobile or aircraft body that was less than half the weight of a comparable metal structure, while offering great simplification of manufacturing and significantly greater crash resistance. Carbon fibre will ultimately find many applications, such as the development of electric motors that weigh less than twenty five percent of comparable internal combustion (IC) engines, but with the same power output.

One particularly fascinating area of materials science was one area of carbon chemistry concerned with carbon nanotubes. After years of research, a number of compelling applications beckoned. For example, a physicist at Wake Forest University developed a fabric made of nanotubes called Power Felt that was capable of generating an electric voltage just from the temperature difference between the warm inside of the glove in contact with the hand and the cooler exterior. The glove could be used to charge a host of small electronic devices such as cell phones.

Researchers discovered a type of polyurethane material that could "heal" damaged sections when it was exposed to UV light. When incorporated into paints, for example, it would mean that cars could fix their own scratches in less than an hour, simply by exposing the damaged area to ordinary sunlight. The long term goal of the research was to develop a whole new world of self-repairing coatings, prosthetics and machines. (Alphonso 2009)

The development of modular Just-On-Order-Machine-Systems allowed almost anyone to develop robots for virtually any type of custom application. For example, simple modules known as Cubelets allowed a complete robot to be created with an unlimited number of configurations. Each Cubelet was just over 1.6 inches (4 cm) in size and consisted of an 8 MHz processor that was programmed to carry out only one specific function. There were rechargeable Lithium-ion battery, rubber wheel, light detection, temperature measurement, Bluetooth and voice synthesizer Cubelets. The modules could be snapped together, held by magnets and they used copper contacts for communication between the cubes. What the final robot was capable of doing depended on the configuration in which the Cubelets were assembled by the user. (Gardiner 2012)

—

The latest aircraft carrier in the U.S. Fleet was named the *Gerald Ford*. It was rumoured to have cost more than $10 billion to build. Clearly such massive outlays for surface vessels were unsustainable and the Navy had to desperately search for alternative ways to build its ships. The solution chosen was to vastly expand the scale of the 3D printers that were commonly used to make spare parts while a carrier task force was deployed at sea, so that they could be employed to make entire ships, even ones as big and complex as aircraft carriers. The technique would be to begin at the bow and to gradually move aft, manufacturing all of the components of the ship including mechanical and electrical systems as the printer moved backwards. The whole ship could be produced in one pass, from bow to stern, with vast reductions in time, effort and expense. The name of this effort could be termed "Just-On-Order-Military" and it would eventually come to dominate the production of equipment in all three branches of the military.

In the Air Force, the first generation of Unmanned Aerial Vehicles (UAVs) were remotely controlled by pilots who flew the vehicles from control stations usually located at major air force bases. Since the signals could be transmitted by satellite, the pilot flying a particular UAV could be located literally half a world away from the vehicle they were flying.

Figure 5-1: One possible design for a future heavily armoured aircraft carrier with ultra-low radar cross section and entirely constructed of 3D printed, nano-composite materials. Source: Tinari

Figure 5-2: The X-45C Phantom-Ray UAV built by Boeing. Source: McLain

By the second decade of the 21st century, there were literally hundreds of different types of UAVs designed for every conceivable military mission. This was also the year in which the US Air Force trained more UAV pilots than pilots for manned aircraft. As an example, the X-45C (The "X" meant experimental aircraft) produced by the Boeing Phantom Works was a test bed for a broad spectrum of advanced technologies including low observability/stealth, active radar jamming, electronic warfare, autonomous operations and high speed operations. The Demon UAV made by BAE in the UK represented a flying wing without flaps and with few moving parts. Instead of using traditional movable flaps, directional thrusters were used to control the craft. The Vulture developed by Lockheed Martin was designed to loiter at above 55,000 ft for periods as long as 5 years for long-term surveillance applications.

The near future will witness the deployment of next generation hypersonic UAVs that will redefine the battlefield. What will enable the development of a Just-On-Order-Air Force will be the application of 3D printing and rapid prototyping technologies to the manufacturing of fleets of high performance UAVs. They would only be built as they were needed on the battlefield. In fact, the capability will be developed to produce modular one-of-a-kind UAVs with unique configurations and properties required for each specific mission. After completion of their missions, the UAVs could be disassembled and the materials recycled into the creation of new UAVs with completely different capabilities.

The X-47B represented the second generation of Autonomous UAVs (AUAVs) where the pilots were taken completely out of the loop and the aircraft was solely under the control of the on-board computers. It required millions of dollars to train a pilot to fly a conventional manned military aircraft. The cost was significantly reduced to train a UAV pilot, and taking into consideration the fact that a UAV did not have to include all of the expensive features necessary to keep the pilot alive (such as life support and ejection seats), this form of aerial warfare was far more cost effective.

The second generation AUAVs dispensed not only with the on-board pilot, but also with any pilot control whatsoever, driving costs still lower. The great advantage with AUAVs was that everything that needed to be known to fly the aircraft under any conceivable conditions was contained in the software stored inside the on-board computers. The "training" to fly the craft was all contained in the development costs of the software. Once the software was successfully developed and fully tested, it could be instantly transferred to an unlimited number of other aircraft at virtually no cost. (Hennigan 2012)

Figure 5-3: The X-47B experimental UAV, coming soon to a war near you. Source: Smevog

In the future, aircraft will get significantly faster and more capable. Air forces around the world were spending huge amounts conducting research to extend the flight envelopes of hypersonic (much faster than the speed of sound) aircraft. The ultimate goal will be to be able to use 3D making technologies to produce every component of a craft capable of taking off from a normal runway and then accelerating to orbital velocity. The ability to accomplish this will result in a dramatic reduction in the costs to get materials into low earth orbit. Below is a model of the experimental X-43A hypersonic craft that was designed to eventually exceed Mack 7 (Seven time the speed of sound). Its performance was analyzed using a CFD simulation in a virtual wind tunnel.

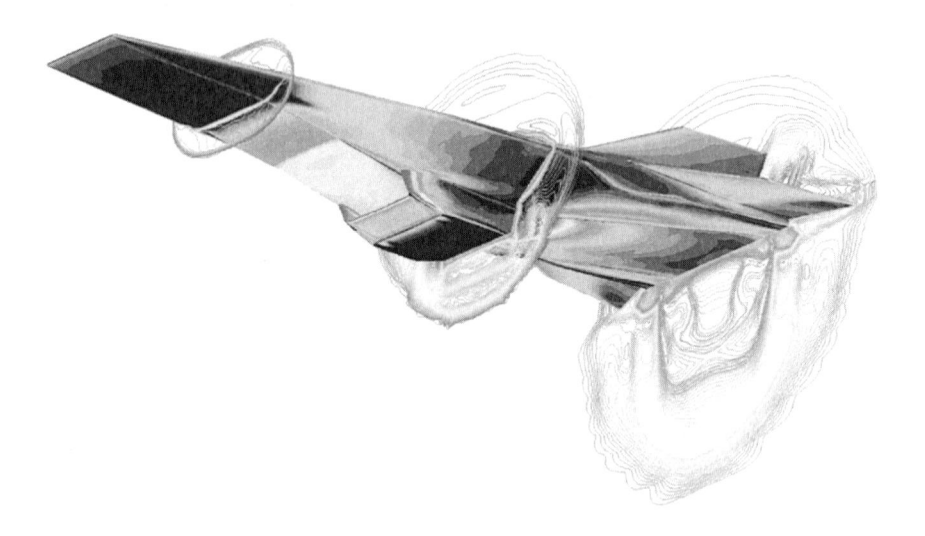

Figure 5-4: The X-43A experimental hypersonic aircraft that will eventually lead to craft capable of affordably flying from earth to orbit. Source: NASA

In parallel with the rapid developments in software, rapid fabrication manufacturing technologies were increasingly being used to build the aircraft themselves as well as all of their component sub-systems. The goal was to have the entire vehicle 3D printed, thereby eliminating all outsourcing of parts and subsystems. The whole vehicle could be manufactured autonomously without human oversight in one room, by a series of 3D printers. The only requirement would be that the printers had to be fed a continuous supply of raw materials and power. This would greatly accelerate the creative process, since every part and system could be continuously upgraded within the computer, and then immediately translated into a real world assembly that embodied the innovative insights of the designer. These achievements of both hardware and software represented the ability to create a Just-On-Order-Air-Force that could put a virtually unlimited number of platforms into the air.

The same technologies that were used to create AUAVs were rapidly being copied by all of the other branches of the military. For the Navy, Just-On-Order-Submarines could be assigned tasks that would be considered far too dangerous for manned subs and for the army, Just-On-Order-Land-Vehicles such as baggage carrying robot-mules, were becoming the grunt's new best friend.

In the near future, 3D making will penetrate to virtually every unit in the military, offering custom designed systems of all types, instant spare parts and replacement equipment as needed by rapidly evolving battlefield conditions.

Significant advances in electronics miniaturization, propulsion and sensors allowed the development and deployment of new generations of Micro-Unmanned Aerial Vehicles (MUAVs). These machines enabled the manifestation of not only Just-On-Order-Military but also of Just-On-Order-Monitoring, a significant component of what has been generally known as the Surveillance Society. This would mean that in principle, every citizen could be placed under observation virtually all the time. Some MUAVs were so small that they could be carried in an ordinary backpack and deployed to make aerial observations almost anywhere. They could be equipped with sensors tuned to detect almost any parameter of interest including visible light, heat signatures or chemical emissions.

These UAVs had military, law enforcement, wildlife, natural resource and engineered structure (such as power lines) monitoring applications. The incorporation of solar cells into these devices could endow them with virtually unlimited endurance. While these machines were impressive enough, next generations of UAVs will shrink still further until they were about the same size as insects, and eventually, to the size of bacteria or even viruses. The applications that such Just-On-Time-Monitoring could be put to were limited only by the imagination of the user. (Globe & Mail 2012)

At the same time there was a convergence of Just-On-Order-Monitoring with Just-On-Order-Manufacturing as increasing numbers of the parts necessary to construct UAV's were produced by 3D printers. Eventually, the 3D making systems located on each military base and on each ship of the fleet would be capable of manufacturing any spare part, component, sub-system, assembly or complete UAV, as required by circumstance.

Rapid advances in technology allowed the development of Just-On-Order-Micro-Motors. These clever devices etched on silicon and powered by hydrogen peroxide could be used to keep new generations of mini-satellites in their proper orbits. The large, expensive, multi-ton surveillance satellites of the past were gradually being replaced by swarms of hundreds or even thousands of cheap, disposable mini-satellites that formed a highly dynamic and adaptable distributed architecture of "sensor nets" around the globe.

—

Anyone who has watched a program such as *MASH* got some idea of the challenges of performing surgery under battlefield conditions. However, as with many Hollywood portrayals of real life situations, the show greatly underestimated the actual chaos in the surgeon's environment. In reality, for real battlefield doctors it was incredibly difficult to keep all of the required instruments in stock, and when they did have all the right stuff, keeping everything sterile enough to actually use was a significant problem.

Finding a solution to this problem was considered to be so important that the Defense Advanced Research Projects Agency (DARPA) launched a special project whose goal was to develop the ability to produce Just-On-Order-Medical-Instruments on the battlefield. Within less than three months, medical modelling engineers had created basic 3D models of all of the instruments commonly used in battlefield clinics. A system was also designed that allowed each surgeon to carry with them the 3D designs for all of the instruments that they had customized for their own use, and as required, these specialized instruments could be manufactured on-site by field-toughened 3D printers.

The first generation of surgical tools were printed out of a very hard, durable plastic which surprisingly, could be given a cutting edge that was sharp enough for cutting human tissues. However, the greatest advantage with digital instruments was that they allowed surgeons the opportunity to provide design input that allowed the tools of their trade to be individually customized as needed. It was then possible to produce the second generation of instruments in traditional stainless steel, or even in more exotic materials such as obsidian (which was able to hold a much sharper cutting edge than steel).

—

The rapid increase in the capabilities of wireless sensors, combined with their continuously decreasing dimensions and costs, led to the inauguration of an era of universal Just-On-Order-Maintenance. Microscopic sensors could now be incorporated into structures such as bridges and high-rises during construction, and they could continuously monitor the stress levels at various points in the structure so that warnings could be issued, giving forewarning that a failure was about to occur. Widespread deployment of such sensors would mean that expensive maintenance programs would be undertaken only when required, saving billions in dollars in unnecessary expenditures. Bio-sensing devices could be incorporated into the human body to warn of impending liver, kidney or heart failure. Wireless brain sensors could even warn of impending mental illness. By giving patients lifesaving pre-emptive treatment, costs to health care systems could be significantly reduced.

Intel and Kraft launched a new generation of vending machines that offered customers Just-On-Order-Marketing. The ultimate aim of the initiative was to offer clients trials of new food products. However, what was different in this case was that the new intelligent vending machine had a camera hooked into a computer that allowed the machine to identify the approximate age, gender and socio-economic group of the prospective customer standing in front of the machine. The device was able to tailor the product dispensed to the identified characteristics of the customer.

Schiphol Airport in Amsterdam installed a machine that was capable of printing out Just-On-Order-Messages. While family and friends were waiting for the arrival of loved ones, they could enter an appropriate welcoming message into the keypad of the machine and it would be printed out onto water-proof canvas as they watched. BannerXpress, the company that conceived and developed the instant banner machines, experienced such an excellent response to the first beta-test machines, that it planned to install machines in airports all over the world.

—

The domain of mineral and resource extraction was profoundly affected by Just-On-Order-Mining. The first breakthrough involved the development of ore-body characterization software that permitted the generation of accurate 3D images of the underground mineral deposit. Companies such as Gemcom developed software packages that vastly improved the economics of exploiting complex subsurface ore bodies. The next major innovation of Just-On-Order-Mining was the introduction of remote controlled mining equipment, meaning heavy trucks and drills operated wirelessly by human operators stationed in air-conditioned control centres located thousands of miles away.

The ultimate achievement of Just-On-Order-Mining will be the introduction of totally autonomous vehicles that can conduct all the mining operations themselves, without any human intervention. In the near future, robots will scour the world for new ore deposits, autonomous machines will exploit the deposit, and then after the resource was exhausted, other machines will rehabilitate the site back to its original appearance. Even the blind could see that JOOM was a major disruptive technology that will soon shake the entire mining industry to the depths of its foundations. (Bouw 2012)

—

The concept of Just-On-Order-Money has long been a staple of Asian culture. Japanese and Koreans were used to paying for all types of every-day purchases with their cell phones. Many Canadian and American consumers were familiar with the ritual of paying for a purchase by simply tapping their payment card on the top of a terminal. Some North American stores introduced an app that integrated a gift or value storage card into the phone so that payments could be made at the counter electronically. The two-step procedure involved download-ing the appropriate app from the company's site and the customer then loaded cash onto their e-card. Each card had a unique QR code. When it came time to pay for a product or service, the client simply launched the app to retrieve the stored QR code, which was then scanned by the clerk and then used to transfer the payment to the merchant.

Many observers were uneasy about these developments because, for example, it meant that a thief could just as easily tap a stolen card as could its rightful owner. Retailers were often delighted with this new "instant digital money" because experience consistently showed that it increased sales. A mother rushed into a grocery store after work to buy some supplies for supper. Her smart phone with active geo-location vibrated and informed her that the store next door was having a "10 min" sale on her favourite summer dresses. She only expected to be buying something for her children's supper, but she ended up spending much more than she intended that evening. And so it goes.

In 2011 a new app was developed called Google Wallet that allowed Android smart phones to be used for payments by just waving them over a terminal. The phone could be programmed to deduct the payment from an existing Master Card or pre-paid debit card account. Security was assured by making sure that payments could only be made when the screen was activated so that money could not be extracted from the device when it was stored in a user's pocket. Also, the user had to be logged into the secure app using a PIN, limiting access to the chip located in the phone. (Iozzio 2011)

All of the technologies already existed to enable Super-Smart Cards that would keep track of all of an owner's assets including jewellery, art, RRSPs, real estate and mutual funds. Purchases could be made on the card and then deducted from the appraised value of the owner's listed assets.

—

Most people were initially shocked to hear of the concept of Just-On-Order-Meat. An organization in Toronto, Canada known as New Harvest sought alternatives to meat harvested from living animals. The first step was to perform a biopsy to extract cells from a living animal, induce them to become pluripotent stem cells, and then to grow these cells in an appropriate nutrient solution. These cells had similar regenerative qualities to embryonic stem cells. The final step was to harvest the resulting tissue and to prepare it just like normally sourced (i.e. slaughterhouse) animal protein.

The first synthetically cultured hamburger was produced under controlled laboratory conditions and tested in the summer of 2013. Large-scale production of meat using this method would offer several advantages over traditional factory farms, most important of which would be that the facility would not be prone to disease or fecal contamination. From an environmental point of view, such edible protein production facilities would substantially reduce the areas of land required for food production in comparison to traditional farming operations, while reducing pollution and increasing the overall efficiency of meat production.

Using animals, up to sixteen pounds of feed were necessary to produce each pound of meat. With synthetic meat, each pound of grain would generate one pound of meat. The challenges remaining included developing a liquid medium capable of supporting the optimum growth of the animal cells and in proving that the process could be economically carried out on industrial scales. However, it was of concern to those working in the industry that this technology would eliminate the need for ranches, feed lots, slaughter houses and processing plants, all significant areas of employment. (Datar 2013) Your JOOM Destiny...

—

The first universities appeared in Europe in the 12th Century, inspired by Islamic centres of learning that were first developed in the Middle East in the 9th and 10th Centuries. Education at the time consisted of students gathering together in a lecture (Latin: "*to read*") theatre to hear an academic read from a hand written book that was far too expensive to be handled directly by the students. The first major threat to the university system appeared in the year 1439 with the invention of the movable type printing press by Gutenberg. Within a few decades, books became affordable by many of the merchant class instead of just the nobility, and there was no longer any need to go into a stuffy room to hear a bearded patriarch read from one. Consequently, in order to survive, universities evolved from just presenting existing knowledge to interpreting knowledge, developing new knowledge and imparting this to students.

Over the centuries, because of their ability to attract world class academics, certain universities gradually gained a significant amount of prestige that allowed them to select the best students for their various academic programs. This meant that certain degrees became more valuable than others, even if the knowledge necessary to get them was mostly identical, just because they had been awarded by a more prestigious institution.

The rise of the Internet as an educational tool permitted the development of a new type of degree, where students could assert that they had mastered certain knowledge and skills. This started the process of separating *what* was known by a particular individual from *where* the knowledge was obtained. In other words, in the future all degrees would be considered equal if a student had demonstrated the required proficiency. Finally, learning would have been decoupled from credentialing. (Yglesias 2011)

The need for researchers to conduct scientific research in well equipped laboratories aided by top calibre graduate students was relatively safe for the time being. This would remain true at least until automated systems reached a level of sophistication where machines were able to make original scientific discoveries on their own. This fact meant that certain post-secondary institutions would be able to maintain greater prestige relative to others, even with the massive democratization of knowledge taking place because of the Internet.

The Boston-based Massachusetts Institute of Technology (MIT) was one of the first post-secondary institutions to put all of its courses on-line and to offer them to anyone who was interested in taking them – for free. The catch was that to actually get the degree, the student had to pay a small fee. But the point was that MIT was leading a trend that could be termed Just-On-Order-Mind, where lectures, courses and entire degree programs were made available through the net to anyone, anywhere, anytime they wanted them.

This and other parallel developments were highly disruptive of the traditional educational system. For example, it was hypothesized that one star calculus teacher could deliver lectures on-line to every grade 12 student in the United States. With the availability of advanced on-line resources such as real-time updated e-textbooks, video simulations and animations and dynamically evolving evaluations that were based on competence and not on the number of hours spent in lectures, the time taken to complete a degree could be reduced by a factor of four. (Wente 2012)

The concept of Just-On-Order-Mind finally came into the traditional classroom in the opening years of the 21st century. According to Nobel Prize winning physicist Carl Wieman, the most important thing to be concerned about in education was what was going on in student's heads. He firmly believed that students brains responded best when they could work together to solve difficult problems and self-organize themselves in teams to solve complex tasks. Using the terminology of social media, he believed that students learned best when they were crowd sourcing. Wieman observed that the traditional lecture did little to develop the skills that would be most useful for the 21st century economy, namely, the ability to work in teams, collaborating, creating, innovating, adapting and synthesizing. Fundamentally, learning was not what the person at the front of the class was doing, it was what the *students* were doing. And if they were just sitting there, they were *not* doing much learning. In the new educational paradigm, the focus shifted from *what* students were learning to *how* they were learning it. Thus the instructor's role had now shifted to helping students find and access information and then helping them to find the best ways to apply it.

—

One of the most frustrating aspects of urban living was the ever present traffic jam. Many highly clever schemes had been tried to address this problem such as allowing cars into the urban core depending on odd/even licence plates, advance parking space reservation, the development of crash avoidance technologies and the use of large-scale car sharing systems. (Ramsey 2012)

One of the ways that technology could address this universal problem would be by the large scale deployment of autonomous vehicles. Cars driven by robotic control would be able to significantly increase the capacity of highways by safely driving vehicles at high speeds with intervals of only a few inches, something that even the best human drivers would never be able to accomplish.

The road to practical Just-On-Order-Mobility was a long and difficult one. To bring the virtually infinite power of private-sector human creativity to bear on the problem, DARPA sponsored an annual contest for robotic vehicles where a significant cash prize would be offered to the entry that could successfully navigate a difficult desert course and cross the finish line first. In the early years of the completion, even the best submissions went only a few feet past the starting line before crashing. But each year, the distances traversed by the vehicles gradually increased until finally, the majority of the vehicles were successfully finishing the race. Soon, it became common for beta-test versions of fully autonomous vehicles to drive themselves from one side of a major city such as Los Angeles to the other at rush hour, without experiencing any adverse incidents. A human did go along for the ride, ready to intervene just in case something did go wrong.

From an informal comment inserted into a presentation delivered to NASA senior administrators in 2009, the author made reference to 3D printing being a technology that they should perhaps be investigating with more vigour. Within less than four years, NASA saw this technology as a central feature of the Just-On-Order-Mobility required for its future planetary exploratory missions. By 2013 NASA's experimental Space Exploration Rover had more than seventy of its parts produced by rapid fabrication including the camera mounts, doors, bumpers and various fixtures. In fact, some of the parts produced had shapes that were so complex and convoluted, that they could not have been economically produced by any other method. The first generation of parts were produced in ABS plastic, PCABS and polycarbonate materials. In the future, parts could be made of virtually any material including stainless steel and titanium. The long-term vision for planetary missions will be to ship the astronauts and the 3D printers to the target planet, and then using locally available raw materials, to print out all of the parts necessary to fabricate the base station and the planetary exploration rover. This approach using Just-On-Order-Mobility would significantly reduce the costs of space exploration. (NASA 2013)

An exciting new area of manufacturing was the creation of Just-On-Order-Mobility with new generations of prosthetics. Designed to mimic the shape of a cheetah's legs and using advanced carbon fibre and other new materials, these devices had given those with disabilities a mobility that had begun to match and that would eventually exceed that of the able-bodied. The key shift occurred when it was decided to abandon the specification that a prosthetic limb should as exactly as possible, look like the damaged limb it was replacing.

With this fundamental mind shift, there was suddenly virtually no limit to the ultimate performance characteristics of prosthetics. A science that evolved from the rapid evolution of prosthetic design was proprioception, the capacity to sense where the body and its various components were in space at any point in time. This was accomplished by embedding computational capacity into the various parts of prosthetics. This would mean that a synthetic foot, for example, could continuously adjust itself to the changes of terrain encountered by the user. It has recently become apparent that the next shift would occur as prosthetics moved from helping to give fundamental mobility to the handicapped, to *augmenting* the mobility of all those who wish it. (Scott 2008)

In cities throughout North America a manifestation of Just-On-Order-Mobility was seen on the streets in the form the Zipcar. The idea was to provide all carless city dwellers with a cheap and convenient form of urban transportation. The user first made a reservation on the Zipcar Web site using a tablet computer or a smart phone. The information was then wirelessly transmitted to the selected vehicle. The locks of the car were released by use of a smart Zipcard and the user would find the keys inside. Refueling of the vehicle could be done for free using the Zip Gas Card, and it was possible to extend the originally planned rental by sending a simple text to Zipcar. The initial success of the concept of short-term Just-On-Order-Rentals meant that such systems were sure to spread to other cities around the world. (Keegan 2009)

—

The development of so-called Intelligent Reactive Clothing (IRC) led to a new era of Just-On-Order-Monitoring. Combining the formally separate fields of textile physics/chemistry, IT, wireless communications, nano-technology and micro-power generation, according to Filiz Klassen, an associate professor of Interior Design at Ryerson University in Toronto:

> "*Reactive garments or fabrics have integrated technologies that make them react to various stimuli, be it light, touch or heat...Materials can be sensitive and change colour in reaction to heat, UV light and electrical energy and others integrate phase-change technology.*" (Jordan 2008)

Indeed, it became apparent a number of years ago that the interaction between humans, their clothing and the environment was going to be one of the next big challenges faced by the textile industry. The first generation of such clothing incorporated superior abilities to keep the wearer warm, cool, dry and unburned, all while keeping its original shape. Second generation clothing did much, much more than just these passive activities. As an example, a snow suit with incorporated electronics including Light Emitting Diodes (LEDs), thin-film solar cells and energy storage batteries, allowed the wearer to not only see while skiing down a dark slope, but also allowed other skiers to see them as well. An intelligent glove was designed to instruct the user on how to properly administer CPR to a patient who was having a heart attack, guiding the wearer to apply just the optimum pressure, frequency and angle of chest compressions.

—

The widespread adoption of so-called "smart" phones equipped with high resolution video cameras was a major contribution to the development of what could be termed the "Just-On-Order-Monitoring-Nation." Now when police officers used excessive force to subdue non-violent suspects, dozens of video cameras were instantly available to record the incident. The prevalence of smart phones with cameras meant that news programs could regularly supplement the material actually captured by their own professional journalists with video material captured by a national network of amateurs that often made up for its generally poor quality with the value of its real-time spontaneity.

Britain spent more than $25 billion to set up a comprehensive system that could monitor every telephone call, e-mail, text message and the browsing records of every person in the entire country. All of this information was dutifully gathered, categorized and stored in a gigantic national data base. The purpose of this huge outlay of both money and resources was the desire to intercept terrorist plots before they could carried out. The City of New York implemented a Just-On-Order-Monitoring-System known (rather ironically) as the *Manhattan Project*. Once again, the main goal was to make it as difficult as possible to launch a successful terrorist attack within the city. Included in the comprehensive program was the setting up a vast network of more than 3,000 video cameras, construction of a government-only wireless network and installation of intelligent cameras, intrusion detectors and perimeter sensors in the subways and other critical areas. The new surveillance system was programmed to recognize suspicious behaviour and to be able to recognize individual faces, then to track the suspects throughout the network. (Shachtman 2008)

Companies that were in the business of insuring private automobiles have long wanted to be able to monitor each of their client's driving habits so that they could more accurately set each person's insurance premiums to reflect the actual risk that they would be in an accident. The advent of a broad spectrum of low cost, miniaturized, wireless sensors meant that virtually any parameter could be monitored in real time including speed, acceleration, reaction time, g-forces resulting from sharp turns and so on. It was also possible to install air sampling sensors that could periodically analyze the driver's breath, to determine if any contentious, behaviour modifying substances had been consumed.

The general name that has been given to this technology was telematics, meaning "measuring from a distance" and it would allow each driver to be issued a totally customized bill. The sensor data was processed by software that also took in GPS information so that at any time it was possible to determine if a particular automobile was travelling faster than the posted speed limit on a particular stretch of road. By providing weekly and monthly reports on their driving habits, drivers could get the feedback that they needed to improve. This system represented another application of "big-data" analytics, where powerful computers were used to analyze huge collections of data to uncover hidden trends. With this approach to Just-On-Order-Monitoring, average premiums charged to huge sub-sets of customers were replaced by individualized premiums custom calculated for each individual driver. (El Akkad 2013)

The first generations of sensors had to be laboriously installed on each individual vehicle. Later, smart-phone apps were developed which tapped into the sensors already embedded into the device to wirelessly tap into the relevant information and then transmit it in real time to the insurance company's computer. One of the ethical issues that was immediately associated with continuous monitoring was deciding what should be done if some of the sensors detect activities (such as speeding) that were considered criminal by the jurisdiction in question. Should the monitor information be provided to the local police so that the driver could be charged?

In recent years there was an explosion in all types of telematics systems that, for example, used animal or human wearable monitors that could monitor the evolution of parameters as function of time such as the intensity of exercise, blood pressure, heart rate, calorie burn rate and much more. One application of this was to allow the majority of at-risk seniors to remain in their own homes instead of in an impersonal care facility, resulting in great savings in medical monitoring expenses. Another application was to monitor horses during races to know when a particular horse was about to experience a systems failure that could lead to a collapse.

Organizations such as iWatchlife began to market home security systems that could in principle identify everyone who entered into a particular dwelling. These systems could also respond to certain events, such as a party involving minors where the consumption of illegal drugs was taking place. Parents were delighted to finally have a technology that could maintain a continuous monitoring of their homes while teenage children were left on their own. But this also meant that police could continuously monitor houses equipped with the system and then move in to apprehend 420-friendly parents when they lit up a fat doobie at the end of a hard day at the office, an outcome that the adults certainly did not intend when they purchased the system.

What information could be extracted about a person by conducting a detailed analysis of their face? Among other parameters, it was possible to determine race, age, sex, possible national origin, genetic and physical health, attentiveness and much, much more. Significant improvements were being continuously made to the two elements that were necessary for effective facial recognition, namely the software that conducted a detailed analysis of images and the comprehensive data-bases that efficiently stored the information on millions of faces. The result would be that an accurate identification could be made of a face, using only a minimum input of information. In many countries, *Facebook* already offered a feature where all of the individuals that appeared in a picture that was posted to the site could be identified, as long as they also were listed in the person's "friend" list. The worry was that when organizations such as Facebook constructed large databases of biometric identifiers and other databases of names, addresses, places and events, it became very tempting to link the two data sets together and thereby create a significant threat to privacy by enabling a determination of who was doing what, when, where, how and with whom.

In 2011 it was announced that Cisco and other Western companies had been contracted to design and build one of the world's largest and most comprehensive digital surveillance systems in the Chinese city of Chongqinq. The system consisted of more than half a million cameras, networking systems, huge capacity digital storage and powerful mainframe computers capable of running facial recognition and other identity confirming software. Costing more than an estimated $2 billion, observers commented that the true purpose of the system was not to reduce crime as claimed by officials, but to suppress political dissent. (Chao et al. 2011)

A rather new but rapidly evolving and broadly diversifying surveillance technology was the concept of a "*Mote*" (Short for: Re*mote* monitoring technology). At its heart, a mote was nothing more than a highly compact, very low cost, essentially disposable microprocessor whose function was to monitor a sensor. The mote could communicate its observations to the outside world using a wireless link. The power consumption of the mote was deliberately kept low by limiting the range of this link to only a few metres and by keeping the mote turned off until an external signal ordered it to report a measurement. A mote was so efficient that it could be kept running for more than a year on a single AA battery. While they appeared innocent enough, motes were in fact spear-heading a whole Just-On-Order-Monitoring revolution of their own.

Motes could be programmed to monitor vibration levels of machinery, the stress levels in bridge support beams, the integrity of security seals in shipping containers, the quality of the water going to a food processing plant, the temperature and viscosity of oil in a pipeline, or crack formation in railway tracks. In fact, the range of possible applications was virtually limitless. The great power of motes lay not in their individual capabilities, which admittedly, were relatively limited. The real potential of motes lay their ability to automatically link to each other to form large, ad-hoc networks with capabilities that far exceeded that of the component motes.

As an example, thousands of motes could be dropped by aircraft into remote forest areas. Once on the ground, each mote would wake up, send a signal to its immediate neighbours, and then establish a wide-ranging network that could monitor soil conditions, determine fire risk, track animal migrations or used to ascertain the extent of an insect infestation. Each mote was only responsible for monitoring its own individual sensor and then for transmitting the readings to all the neighbouring motes. All of the data collected by the network was funneled, mote-to-mote, to a central collection mote that was equipped with a wireless link that was powerful enough to reach, for instance, up to an overhead UAV.

The advantages with such mote networks included their great flexibility, adaptability and their ability to monitor huge areas at very low cost. In the future, motes will continue to shrink in size, increase in capabilities and fall in price. Eventually dust-sized motes will be extensively deployed over wide areas, for the most demanding applications, and with each mote in the network costing less than a dime.

—

Just how far could Just-On-Order-Monitoring go? Researchers in Japan managed to extract images directly from the minds of subjects by the analysis of brain scans. In theory, for the first time this allowed external access to a person's most personal and previously private thoughts. The original aim of the research was to open a new communication pathway for those unable to speak or communicate in any other way. The operation of the visual system was in principle simple. Light entered the eye where it was converted into electrical impulses by the retina. These signals were transmitted to the visual cortex of the brain where the signals were decoded, interpreted and translated into visual images that the subject interpreted as vision. The researchers used a medical scanner to intercept the signals going to the visual cortex, then they employed a complex algorithm to interpret the signals and then converted them into visual images that could be interpreted in the laboratory.

The algorithm was originally calibrated by having volunteers look at hundreds of different images, including letters of the alphabet. A digital data base of the resulting electrical signals was then created. Since the electrical signals of the brain created external EM fields that could be detected (i.e. with a Superconducting Quantum Interference Device (SQUID) device, the next stage of this research could be the development of remote methods to monitor human thoughts. As this technique was perfected, huge ethical issues were raised when the capacity to directly monitor a person's most intimate thoughts could be made available to authorities. Could individuals be jailed for crimes that they only thought about committing? Will hackers be able to extract passwords directly from people's minds as they extract money from their ATMs? (Kubota 2008)

Until recently, researchers believed that each individual had internalized clear moral standards of right and wrong hard wired into their brain from upbringing, education and environment. This assumption was now being seriously questioned. In 2010 researchers discovered that test subjects could have their moral values modified by exposure to varying doses of Transcranial Magnetic Stimulation (TMS). It was found that magnetic stimulation of the brain in the Temporoparietal Junction region (near the right ear) caused subjects to tend to focus on the outcome of actions rather than on the intent. Thus, actions that they had previously seen as morally questionable tended to be ranked as acceptable after a single session of magnetic stimulation. This preliminary research indicated that some day it would be possible to create Just-On-Order-Morality that could be used by unscrupulous governments to manipulate behaviour. (Lametti 2010)

It should also be mentioned that, if desired, TMS can also be used to erase selected memories entirely. The practical possibilities of this have only been slightly explored. A loving spouse whose partner committed suicide in a particularly gory way could choose to have the memory selectively erased. Or more chillingly, governments could choose to assure a uniformly happy citizenry by ordering that everyone have any "unhappy" (i.e. dissident) thoughts selectively erased, under proper medical supervision of course. One of the more frightening aspects of your JOOM destiny.....

—

The rapid fall in price of sensor and wireless technologies led to utilities in many jurisdictions abandoning traditional human electrical "meter-readers" and to replace them with automated "smart" meters that could read electrical consumption minute by minute in real time and then continuously report its findings to a central database located in company headquarters. The system was so refined that not only could it detect illegal marijuana grow operations in a private home, but it could also detect the number of human occupants in non-licensed basement suits. The privacy violations possible with such continuous monitoring technologies were significant but remained mostly unexamined by political representatives.

It was discovered that there was a protein in the hypothalamus of the brain that had an influence on the aging process. Once the protein in question was identified, it became possible to find methods to slow down aging in particular individuals. (Andreatta 2013) Combining this development with ability to print out new human organs from stem cells, the result is what could be called Just-On-Order-Mortality, where thousands of individuals could in theory, live virtually forever, or until they actually made the conscious decision to die (or until impatient great-great-grandchildren made the decision for them).

It was never a secret that the largest global manufacturing concerns had long received political and economic concessions from powerful national decision makers. Obviously, anything that disrupted traditional manufacturing would naturally impact national political structures. Political democracy was, by definition, when all of the important structures of government were under the control of and by the consent of the governed. At one time, all manufacturing was in the home and was under the direct control of the people who directly benefited from the goods being created. The rise of the factory system led to development of first local, then national and ultimately multi-national corporations, trade unions and protective government structures that took the control of manufacturing away from ordinary people.

Fundamental, transformative change could only happen when control of manufacturing was taken away from unelected industrial boards of directors and faceless government bureaucracies, and returned to ordinary people. As shown in this book, the Internet was the critical tool that enabled the democratization of book publishing, broadcasting, journalism, music, video and all other media communications, enabling for the first time the full participation of every citizen who wanted to. The Internet allowed, for the first time in history, the creativity and innovation of ALL people to be tapped for the benefit of society, and not just a few corporately employed industrial engineers, NASA rocket scientists, socially favoured designers, politically correct writers, network approved newscasters or government approved bureaucrats.

A revolution was happening in manufacturing, a fundamental paradigm shift that would eventually affect everyone on the planet. Yet the powers that be were barely cognisant of this fact. This meant the policy makers would be caught by surprise when the social, political and industrial structures that they have cared for and nurtured for so long, suddenly collapsed.

In the first phase of the JOOM revolution, designers and inventers could create a new device concept on their computer and with a single touch of a mouse button, they could send it to be immediately fabricated by a Chinese assembly line. In only a few days, a prototype would appear at their door. With a few more clicks of the mouse, thousands of commercial versions of the device could be manufactured, placed into a standard shipping container, and in a few weeks, shipped anywhere in the world.

The second phase of the JOOM revolution came with the development of systems such as Fab-Lab, that allowed the delivery of micro-factories to virtually anywhere they were needed. Suddenly, entrepreneurs no longer needed large, centrally located factory assembly lines, cutting time and delivery costs by using neighbourhood micro-factories to make prototypes and limited production runs of innovative new products.

The third phase of the JOOM revolution will come with the wide-spread arrival of advanced 3D printers into the home, essentially turning every home office into its own micro-factory. At this point, history will have come full circle: In the 1700s virtually all manufacturing took place in the home or in its immediate environment. The rise of the factory system, with its assembly lines and human regimentation, put manufacturing firmly into the hands of large corporations and investment capitalists. Now, more than 200 years later, technology in the form of affordable CAD software packages, easily accessible design tools, web-based collaboration, Internet distribution and most importantly, 3D printers and CNC machines, now allowed manufacturing to return to the place where it had been for most of recorded history, in the home.

Cory Doctorow summarized this situation in his book, *Makers*,

> "*The days of companies with names like 'General Electric' and 'General Mills' and 'General Motors' are over. The money on the table is like krill: a billion little entrepreneurial opportunities that can be discovered and exploited by smart, creative people.*" (Doctorow 2009)

Technology was causing barriers to entry to crumble everywhere. It was no longer necessary to have access to huge amounts of cash or to have high-level political connections to get real, globe changing stuff done. If someone wanted to make their own custom news programs to compete with network news, they could. If someone wanted to create, produce and sell their own music, go for it. If someone wanted to make their own feature length movies, yes!. The *Blair Witch Project*, made with low-cost, amateur equipment and with a production budget of about $22,500 went on to make more than $249 million at the box office. At no other time in history could such an accomplishment have been possible.

Never before have so many sophisticated tools been available to the ordinary person at such a low cost. In the past, only the largest corporate enterprises could afford full sized CNC machines. Now, anyone could have access to one at: www.shopbottools.com or: www.buildyourcnc. com. Home-based designers could now do what only IBM and Intel used to do, namely, design and build their own circuit boards and integrated circuits by use of the tools available at: www.AdvancedCircuits. com. In fact, virtually any item that could be imagined, had a Web site that provided the tools necessary to aid in its manufacture.

It is now time to ask some very important fundamental questions. In the 1930s, economist Ronald Coase asked the innocent question: "Why do firms exist?" The simple answer was: "To minimize transaction costs." But Coase could also see that as firms increased in size and attempted to deliver increasing numbers of products and services, the costs of oversight, in the form of increasing numbers of management bureaucrats and a ballooning in the number of rules and regulations led to the gradual loss of competitiveness of the large firm relative to its much smaller upstart competitors.

Until the 1970s, large firms like IBM specifically made a point of hiring the smartest students graduating with Ph.D.s from the nation's universities every year. Then they began to realize that the major flaw in Coase's argument was that even with all the millions that they had invested in hiring the smartest people, most of the most outstandingly brilliant and innovative people either worked for some other firm, or even more incomprehensively, actually made the choice to work for themselves.

One of the most powerful effects of the Internet was that it greatly facilitated the process of bringing together geographically disparate groups of individuals for a common purpose. What usually held such ad-hoc groups together was that all the members shared a particular passionate vision for what the future should look like. This fact represented a fundamental paradigm shift in the way people could organize themselves to do business. The "leaders" of such dynamic groups did not see themselves as presidents or CEOs, but as facilitators, co-ordinators or motivators. The highly talented individuals working in such groups did not wait to be assigned to a particular aspect of a project, but tended to contribute to the project in any way that they thought would be most expedient. The contributions made by all group members were welcomed, even if a particular member had only one innovative idea to offer to the project.

As an example of a successful product developed by a loosely connected community of developers mostly free of traditional hierarchies, chains of command or rules of procedure, consider the Internet browser Firefox. The software was developed in record time and it rapidly went on to challenge Microsoft's Internet Explorer. Other examples of successful open-source products include Linux and Wikipedia and there were many others.

In the future, the increasing capabilities of the Internet will allow open source development to be extended to many knowledge areas such as medicine, engineering, genomics and others. Just one example of open source successfully being used to raise money for innovative projects was the Kickstarter business model. (Kickstarter.com) Traditional markets worked because of strong financial incentives awarded to market leaders and innovators. Hierarchies worked because those at the top could enforce their directives on those below. So how was it possible for open-source projects to work?

At first glance, it did not seem obvious why hundreds or even thousands of unpaid volunteers would successfully collaborate on a complex design projects and sustain the initiative for months or even years. And then, after all of that uncompensated effort, the participants had to watch as the result of their efforts was given away for free. The fact was that participants in open-source development projects often stated that they took part just to experience the joy of contributing to a leading-edge development project.

An open-source innovation was public and non-proprietary, anyone could take the design, build it and then use it for any purpose to research how it worked, use it for any purpose, improve the design and then share the modifications with all those in the "crowd" so that all could learn and benefit so that they could in turn make their own improvements to the design.

The three essential features of open-source innovations were:

- The innovation must be made available to any interested party for no more than the cost of distribution

- Anyone could redistribute the innovation for free, without royalties or licensing fees to the original designer

- Anyone could modify the design or generate new designs from it, and then distribute the modified design under the same terms.

Talented individuals were most likely to participate in an open-source development project if each of the contributors:

- Could clearly judge and evaluate the viability of the evolving project

- Had all of the information and feedback necessary to make an informed evaluation of the results of their contribution

- Were driven by personal motives that went beyond simple economic gain

- Could continuously gain valuable knowledge and experience from their contributions to the project

- Held a firm belief that the project was ethical and was making a positive contribution to the world. (Weber 2004)

Summary of the Most Significant Trends Associated with JOOM

It was the purpose of this brief work to outline the basics of the global JOOM revolution and to describe some of the many ways that it was transforming your world. Some visionaries have stated that 3D printing and the associated technologies that have been examined here will transform society as fundamentally as the discovery of printing or the development of the first practical steam engines. This trend will lead to fundamental changes in all aspects of our world including economics, engineering, medicine, media, sociology, human relations and entertainment.

Trend 1: *The traditional large industrial firm of hierarchies, defined structure, chains of command, bureaucracies, control, procedures and specified benefits will rapidly give way to entrepreneurial individuals and very small, virtual, ad-hoc, dynamic organizations that will be able to rapidly capitalize on new ideas and will be equally ready to drop ideas that have run their course and have become obsolete.*

Implication: The big three automakers and other large "20ᵗʰ century style" manufacturing entities will give way to smaller regional and eventually neighbourhood custom manufacturers based on rapid prototyping and 3D printing. This implies the complete undermining of the established power structures created by networks of large corporations including big labour unions such as the UAW, the seemingly endless streams of taxpayer funded financial support from political policy makers and the influence that corporate executives exerted on the political process and on the directions of technological innovation in society. In a nutshell, *creativity will be democratised.*

Trend 2: *Manufacturing of low cost goods will rapidly shift from the assembly lines of China and other developing nations back to the homes and neighbourhoods of North America.*

Implication: The post-World War II economies of Germany, Japan, South Korea, and China all largely based the achievement of their national development goals on the aggressive exporting of low cost manufactured goods. The setting up of factories geared to making low-cost products for export was almost considered axiomatic to national development. With the rise of highly versatile, home-based, North American manufacturing, future nations will not be able to launch their development by focusing on export-led growth. At the same time, the thousands of factories built with low cost international markets in mind will suddenly find the ground falling from under their feet and will be forced to carry out a serious re-evaluation of their existing business model, or they will have to close. In a centrally planned economy in a nation such as China, this could lead to intolerable social strains as millions of workers who aspired to enter the middle classes suddenly find themselves unemployed and without a social safety net.

Trend 3: *As large-scale, low cost manufacturing goes into decline around the world, patterns of international trade will be forced to change dramatically.*

Implications: After the global economic collapse of 2008, there were a number of times when freight rates for containers going from Asia to Europe fell to zero. The difference between this collapse and previous recessions was that the industry did not see this as a regular cyclic slowdown of economic activity but rather as a long-term, tectonic shift in demand. Someone going down to any major port such Seattle, WA, or Vancouver, BC was likely to see thousands of containers awaiting transport by truck or train to retailers across the continent. Many of those containers were holding low cost manufactured goods made on assembly lines in the developing world.

Container traffic will have to undergo a dramatic reduction when low-cost manufacturing shifts back to North America. At the same time, as 3D printers continue to fall in cost while also increasing their capabilities, home and neighbourhood manufacturers will continuously increase the quality of the products that they will be able to produce, affecting increasing numbers of large manufacturing concerns around the world. The gradual transition from daily printed newspapers to on-line digital downloads has already undermined the viability of the pulp, paper and the newsprint business and affected all those who earned their living in related industries such as "beachcombing." While most video rental and music stores were forced to close or dramatically restructure because they simply could not cope with the efficiency of Internet downloads, similarly, many traditional bookstores discovered that for the first time they could not pay their monthly rent, as people increasingly ordered their books from Amazon, or even more efficiently, by-passed paper altogether and directly downloaded the e-book to their digital reader.

Trend 4: *Small, informal, modular and mostly virtual firms equipped with the latest in custom enhanced 3D printing technologies will accomplish what the large firms could not, such as, for example, putting payloads into orbit at very low cost, or making an affordable electric car with a range and cost that will make it attractive to the average suburban buyer.*

Implications: Electric vehicles are real game changers, this is why the large firms have kept their prices so high that the average driver would never be interested in buying one. When creative entrepreneurs begin making all of the components for a low-cost electric vehicle that will actually sell, dozens of billion dollar industries employing hundreds of thousands of wage earners will be doomed to oblivion. Electric cars do not need gasoline/fuel pumps, oil, transmissions. radiators/cooling systems/fans, mufflers/exhaust systems, spark plugs and much more. Each of these represents a major industry doomed to destruction, and there is little that the large firms can do to stop it. At the same time, fuel taxes represent a major revenue source for governments around the world. Electric vehicles do not use fossil fuel, so governments will be hard-pressed to replace what was a lucrative revenue source. It will prove difficult to selectively tax the electricity that is used to charge an electric vehicle and not the power that is used to light the garage that it is stored in.

Trend 5: *3D printing will increasingly do what before, only certified professionals could do.*

Implications: You accidentally sat on your prescription glasses breaking a lens. Dang. You went to your optometrist, got your prescription, ordered a new lens and then waited for the new lens to be ground and polished. After all that, you were charged $698 for your trouble. In the new paradigm, you sit on your glasses and break a lens. Dang. You go to a 3D Web site and go to the lens section. There you enter your prescription and you proceed to download the design of the lens to your tablet. You then print out the lens in polycarbonate and within 35 minutes you have your new lens.

As another example, professionally made orthotics designed to correct a number of biomechanical problems can easily cost up to $1,000. It will soon be possible to make a 3D scan of the patient's foot, take a video of the person's walking gait, run the scan and the video through a dedicated orthotics design software analysis and then generate a design in STL format that can be manufactured by a 3D printer. Total cost: $0.99. Technology will soon allow hundreds of thousands of professionals in hundreds of different fields to be by-passed in this manner, leading to massive social dislocation.

In the long-term, increasingly sophisticated design software coupled with the next generations of 3D printers will evolve to the point where ordinary citizens will be able to build and deploy virtually anything, including their own advanced UAVs, subs, land vehicles, artillery and other advanced military weaponry. Regulators will have their hands full attempting to stem this tsunami of technology

Trend 6: *The 3D printing of other 3D printers and of fully autonomous robots will lead to the age of Just-On-Order-Machines that will increasingly do the formally well-paying jobs that were too difficult and/or hazardous for humans to do such as radioactive waste disposal, deep earth mining and jungle exploitation.*

Implications: As autonomous machines increasingly create other fully autonomous machines, the cost of robots will continue to fall, allowing them to spread until they gradually become ubiquitous. The question can be asked: what will humans do when most of the things that kept them occupied are done by machines that can do the same things faster, more accurately and with significantly less downtime than any human. Will we see modern Luddites breaking into the new manufacturing centres and smashing the machines that undermined their ability to earn a decent living?

Trend 7: *The massive introduction of all types of autonomous vehicles will lead to universal Just-On-Order-Mobility.*

Implications: Drivers will be able to walk out to their private vehicle and program in their desired destination and then just sit back and enjoy the ride. Dramatic changes are in store for the economy as increasingly, truck, bus and limousine drivers are gradually phased out and replaced by fully autonomous vehicles will roam the nation's highways without human intervention. The capacities of streets and highways will be significantly increased as speed limits are gradually raised for robot controlled cars and trucks (but not for those still driven by humans) because the artificially controlled vehicles will be able to safely drive only centimetres apart. No longer limited by driver endurance, trucks will be able to cover long-haul routes non-stop, bringing fresh produce to market sooner and at dramatically lower cost. When equipped with advanced weather and road sensors, vehicles can be made to drive at the ideal velocity for each set of prevailing conditions, so that fuel consumption and vehicle depreciation can be minimized.

Trend 8: Because of *Just-On-Order-Monitoring, more than ever before the lives of private citizens will be visible to others, including government agencies, security services, hosts of websites, shop owners and to each other.*

Implications: Privacy will become an increasingly rare commodity as virtually every occurrence will be recorded and immediately posted on-line. With the rapid spread of video and Internet-equipped smart phones, virtually every citizen acquired the capacity to provide real-time monitoring of everything that happened in their immediate environment. Combined with the proliferation of fixed mounted surveillance cameras and cheap, small and accurate sensors, the result was the creation of an essentially universal Just-On-Order-Monitoring capability and the birth of the comprehensive surveillance society. Functional democracy requires transparency in government. Instead, Just-On-Order-Monitoring has ironically led to each citizen becoming transparent to increasing secretive governments.

Trend 9: *The organizations that will be the most successful in the new economy will be those that will be able to efficiently harness the creativity and intelligence of not only their own work force, but also of their customers. This will be accomplished largely by transforming the structures of the organization from traditional command and control hierarchies to creative and dynamic networks of participants.*

Implications: The kind of creativity that used to take place within the laboratories and design studios of large, well-financed organizations will increasingly occur across the self-organized networks linking together entrepreneurs. Things in the world will increasingly evolve from just being "items" to being *congealed information*. The personal customization of all products and services will be vastly enhanced by the storage of information in the infinite Cloud, instead of within the limited item itself. In this world, the maintenance of a high degree of connectivity will become *the* factor of utmost importance.

Trend 10: *The war over who will control the future means of production in the world will be a long and bitter one, but eventually those attempting to restrict intellectual property and to stifle human creativity will lose.*

Implications: The privacy battles now being fought over the data files created by 3D scanners and the products produced by 3D printers have only begun. In 2013 *Autodesk* released a free iPhone app called 123D Catch that allowed a user to scan any object/person they encountered. While the app was primitive and didn't work very well, the technology was only an indication of even greater capabilities that would soon be placed into the hands of ordinary people. The time was soon coming when home manufacturing systems centered on 3D printers could literally produce anything from slippers and new chemical compounds to complete functional automobiles.

The holders of intellectual property rights struck back with a number of initiatives, one of which was setting up Intellectual Ventures Inc., a company run by Nathan Myhrvold, the former CTO of Microsoft. The company patented a system that would gain universal control of the production of any physical object by requiring that any 3D printer could only produce an object after the file had been validated against a database of items authorized to be produced. It was hard to describe this initiative as anything but cumbersome and creativity killing. Apple attempted to impose Digital Rights Management when it set up iTunes, but abandoned the attempt when it became apparent that it would stifle the growth and profitability of the site. The universal rule that applied here was that information just wanted to be free and that those who tried to enslave it would eventually end up on the ash-heap of history. (Mitchell 2013)

Trend 11: *Customers will play an increasingly important role in all aspects of product and service development.*

Implications: Just as quantum mechanics required a revision of the view of subatomic matter as "particle-waves," the new Just-On-Order-Marketplace will no longer see consumers as passive recipients of independently designed products, but rather, as active "consumer-innovators" who will increasingly take active roles in the design and development of new products. There were numerous benefits to corporations from this development, including the fact that contributing consumers rarely expressed concerns about the control of the intellectual property of the innovations that they helped introduce, so organizations that could succeed in tapping the creativity of their clients could essentially do this with little additional costs to themselves. The fundamental question that each organization had to ask was how they could facilitate this highly beneficial process. They will have to examine every aspect of their operations and ask what changes have to be made so that consumers can more efficiently develop, prototype, beta-test, and market-test every innovative new product being conceived by the organization's professional product development team.

Trend 12: *Technology will have a significant impact on the all sectors of society, but especially on the middle class.*

Implications: Technology has dramatically changed business innovation in three ways: Stage, Complimentary and E-Commerce. Companies such as Amazon offered thousands of other organizations a stage or platform to sell their products or services. The second item on the list referred to the offering of items to customers without monetary charge. The zero costs associated with the distribution of information through the Internet meant that an incredible assortment of things could be offered for free. For example, books whose copyright had expired were made available for free on the Internet. The third item in the list refers to the fact that in the near future, more than fifty percent of retail sales will be done on-line. According to Yuri Milner, a Russian Internet investor, the combined impacts of these three trends will be the loss of more than 40 million, mostly middle class jobs in the next twenty years. It is not certain if replacements for these jobs, if any, will be of the same quality.

Trend 13: *The continuing decline in the both the size and cost of sensor technologies combined with the increased computational power of smart phones will lead to the establishment of increasingly effective sensor networks for diverse applications.*

Implications: Research conducted since 2010 led to the development of sensors that could be combined with smart phones to provide a coverage area that will eventually become global. One group of sensors will detect dangerous industrial gases such as chlorine (Cl), carbon monoxide (CO) and ammonia (NH3) while others could detect harmful biological agents. A porous silicon chip in the sensor experiences a change in colour when exposed to certain chemical agents and this change can be detected by a miniature camera built into the phone. The camera-computer analyses the change in the sensor and automatically transmits the results to a central computer. The central computer compiles the sampling results of thousands or even millions of sensors, allowing the exact extent of the toxic plume to be established with great accuracy. A call can then be made to every cell phone in the delimited area telling the phone owners to evacuate the area.

Trend 14: *The DIY (Do It Yourself) and Open-Source movements will become increasingly prominent throughout the world.*

Implications: *OPEN* systems will gradually swamp out proprietary business models. OPEN systems development is a central component of the *DIY* movement. Supported by publications such as *Make* magazine and Web sites such as Instructables and iHive3D, and driven by increasingly accessible and affordable scanning technologies, 3D modelling software and online global collaboration, DIY hackers will increasingly engineer, modify and improve their own consumer products. In the past, corporations spent fortunes in market research to find out what their customers wanted. In fact, customers already knew what they wanted, so it was far more efficient to give them the tools to design and manufacture it themselves.

Trend 15: *The advent of a wide variety of increasingly inexpensive 3D printed sensors, implants, actuators and miniaturized systems will speed the convergence of man and machine.*

Implications: Futurist Ray Kurzweil predicted that a time would come when artificial Intelligence (AI) would surpass human intelligence and when machines would merge with human biology to form a new type of superior being. Kurzweil called this happening "The Singularity." When superior intelligence begins to design successive generations of increasingly powerful intelligences, the system will enter an exponential growth phase.

Trend 16: *Just-On-Order-Software or SaaS (Software as a Service) will become increasing significant to organizational operations.*

Implications: SaaS is a delivery method for the most up-to-date software that provides access to its functions remotely, as a Web-based service. Just-On-Order-Software allows organizations to access business functionality at a cost that is typically significantly less than paying for traditional licensed applications since the pricing is based on a monthly fee. Also, because the software is hosted remotely, organizations don't need to invest in any additional hardware. Just-On-Order-Software replaced the need for organizations to handle the installation, set-up, daily upkeep and maintenance. It also eliminated the need for several on-site software related staffing positions.

Trend 17: *3D printers will continue to become less expensive and more capable until a machine costing less than $1,000 will be available to home users that will be able to manufacture virtually any object out of almost any material. One of the trends that will be driving the collapse in prices will be the expiration of many of the patents that have constrained creative innovation in various 3D technologies.*

Implications: Around the Maker world, entrepreneurs were solving the various problems associated with the creation of low-cost 3D printers capable of printing functional objects in stainless steel and other metals. As one example, a company named Metalysis developed a process that began with a naturally occurring Titanium ore (rutile) and converted it into a pure metal powder using electrolysis. This low cost powder could then be used as a feedstock for metal 3D printers. (Metalysis 2013) With the widespread availability of printers capable of making exact replicas of metal components, it will be possible for anyone to cheaply manufacture anything that they may need within the privacy of their own home, such as spare parts for antique vehicles. This capability will eventually make spare-parts supply networks run by large corporations completely redundant. Moreover, individuals for the first time will be able to manufacture their own consumer products such as snow mobiles, bicycles, electric automobiles, aircraft and even small spacecraft. The full consequences of giving such creative power to individuals remain to be explored as part of the JOOM destiny....

Trend 18: *Human tissues, bones and organs produced by 3D printers will become increasingly common, eventually having a dramatic impact on life expectancy. Eventually it may be possible to make most humans essentially immortal.*

Implications: Increasing numbers of people will be able to replace their organs as they fail with virtual identical copies printed from their own stem cells. This will mean that many individuals will begin living to the estimated ultimate human life span of 150 years. On the plus side, seniors will be able to replace diseased or damaged joints so they will be able to remain productively active much longer. On the negative side, it is not known how society will be able to handle a situation where many of its oldest members simply refuse to die.

Trend 19: *3D printers will eventually become as common in the home as ink/laser jet printers are today. At the same time, the capabilities of each subsequent generation of machines will increase until it will be possible for virtually anyone to manufacture essentially anything, anywhere and at any time, and this capability will have profound political impacts.*

Implications: This is an extension of Trend 17, but looking at some of the political consequences of the trend. Some futurists can imagine a new era of "universal abundance" mediated by 3D printers and related technologies. But it can also be imagined that the effectiveness of governments to exert positive control over populations will be severely eroded when, for example, gun control laws are made redundant by the almost universal availability of home manufactured, military grade, assault weapons and ammunition. And this is only the beginning. How will product safety laws be enforced when everyone can make their own versions of products such as snowmobiles and hang gliders in the privacy of their own homes, completely free of regulatory oversight?

In the spring of 2014, a dock workers strike in Vancouver, British Columbia created a severe shortage of spare parts throughout western Canada. Mountain bikers wishing to buy replacement parts for their bikes were told that the components were sitting in ships anchored just off-shore. The ships were uselessly waiting for the strike to end so that they could be unloaded. Essentially, a few hundred unionized workers were using their political power secured under the *ancien régime* to hold millions of their fellow citizens hostage. In the near future, the advent of universal 3D printer ownership will make such exercises in social manipulation useless because everyone will be able to make all of their own spare parts as needed.

Trend 20: *3D printers and supporting technologies will be used to "kick-start" the economies of developing nations leading to an era of more equitable global prosperity.*

Implications: Medical clinics in the most remote jungle outposts and classrooms in the most inaccessible mountain valleys, once equipped with the newest technology, multi-material, 3D printers and high-speed, wireless Internet links will be capable of manufacturing all of the items and equipment that they will need to compete with the best funded and well equipped medical centres and educational institutions anywhere in the major urban centres of the developed world. This will lead to a new era of global wealth and prosperity unprecedented in human history.

Trend 21: *The increasing capabilities of 3D scanners to record and digitize all the details of rooms, buildings, large structures and outdoor* *environments, combined with increasingly powerful software capable of creating and displaying highly detailed and fully interactive virtual worlds, will lead to whole new era of virtual tourism.*

Implications: It is a little known fact that tourism is the world's largest industry. Gradually, all of the globe's major tourist destinations, both past and present, will be fully digitized and posted to the Cloud. Anyone with a virtual reality headset and an internet connection will be able to undertake any touristic exploration that they may desire such as counting the number of stones making up the walls of the King's Chamber in the Great Pyramid (exactly one hundred blocks), taking in the view from the uppermost balcony of Notre Dame Cathedral of Paris, climbing the stairs of the great Lighthouse of Alexandra as it was in the time of Cleopatra, or exploring the hundreds of chambers within the Angkor Wat in Cambodia. While many individuals with sufficient financial resources will still prefer to go to actual real-world travel destinations, millions of people who previously could not have afforded to go to exotic locations, will now have the means to do so in the safety and comfort of their own homes.

Trend 22: *The use of crowdfunding will become increasingly important in the raising of capital to launch new 3D printing and other technology R&D and commercialization ventures, as long as the regulators do not kill it.*

Implications: In March of 2014, the securities commissions in a number of Canadian provinces announced proposed rules to control the raising of capital and the distribution of shares in new ventures by crowdfunding sites. The rules were designed to limit the maximum amounts that could be raised per calendar year and how much a single individual could invest in a given project. All crowdfunding sites would be required to be registered with the securities commissions and would be required to comply with established minimum capital, insurance and reporting requirements. The aim was to reduce the incidence of fraud while at the same time maintaining the proven capability of crowdfunding sites to allow innovators to quickly and efficiently raise the capital needed to launch new ventures. It was hoped that the new regulations would not crush the powerful creative spontaneity of the crowdfunding concept. (Gray 2014)

Trend 23: *The Just-On-Order-Movement will gradually encroach into numerous sectors of society such as employment and corporate governance.*

Implications: An increasingly important trend in corporations was the adoption of Just-On-Order-Employment, also known as Results-Only-Work-Environments (ROWE). In this model, the function of management was reduced to setting the various goals that employees had to attain and the deadlines when the results had to be completed by. After that, it was completely up to workers to plan and make all of the decisions necessary to accomplish the specified task. They could work when and where they liked, work alone or assemble teams as required and tap all of the necessary corporate resources necessary to reach the goal in the established time frame. Early adopters of this model reported that the creativity of workers was significantly enhanced when they were freed from traditional corporate oversight and micromanagement structures and were allowed to do what they did best - innovate. Just like information, workers wanted to be free. They also wanted to be treated like responsible adults.

Trend 24: *The Internet will continue to evolve and will become an increasingly effective and capable enabler for many of the technologies associated with Just-On-Order-Making.*

Implications: The ability to scan a 3D object using a smart phone, compress the resulting data file, transfer the data over the Internet to a distant location equipped with a 3D printer, and then recreate an exact copy of the original object will be a killer app for the 21st century. It is as close as we are likely to get to the Star Trek transporter technology for the foreseeable future. It was one of the central purposes of this book to examine some of the consequences of this amazing capability. But the technologies discussed here will soon look like Model-T's compared to what is coming.

Trend 25: *There will continue to be an increasing decoupling between the increase in national GDP and energy consumption per capita.*

Implications: In the past, each increase in GDP was closely matched by a corresponding increase in the energy consumed per capita in a given society. With the introduction of an increasingly capable internet backbone, web and supporting technologies, the transmission of information using solid media such as paper, film and metal will increasingly be replaced by direct source-to-consumer digital transmission. This will result in dramatic reductions in energy consumption. It was once estimated that each daily issue of the *New York Times*, required the killing of tens of thousands of trees and burning huge volumes of fossil fuels for the distribution of the resulting newspapers. All this will be eliminated by the full digitalization and internet distribution of newspapers, periodicals and books. Similarly, the ability to scan 3D objects and to transmit the digital files through the internet to virtually any point on Earth (or even into interplanetary space) will result in dramatic reductions in manufacturing, shipping and energy costs.

Trend 26: *Computing technologies will continue their inexorable advances resulting in increasing powerful and sophisticated digital capabilities being shoehorned into ever-smaller packages.*

Implications: Smart phones and their future evolutionary derivatives, will soon provide their owners with computational capabilities that only the most powerful supercomputers could boast of a few years ago. Each person will eventually be able to accumulate more than a petabyte (1,000 terabytes) of personal data files and will be able to access more than a yottabyte (equal to 1,000 zettabytes, where 1 zettabyte equals 1,000 exabytes and where 1 exabyte equals 1,000 petabytes) of global information. This will mean that each person will have instant access to the entire store of accumulated human knowledge, anytime, anywhere. It can be hoped that this will lead to an unprecedented explosion in human creativity and innovation. However, it remains to be seen if this is in fact what will actually happen. Only time will tell.

Trend 27: *For the first time in history, people will be able to transcend the dictates of religion, politics and culture and become truly free to make their own choices in life.*

Implications: Millions of girls in the developing world have been forcefully subjected to female genital mutilation for cultural and/or religious reasons. More than a billion boys around the world have been subjected to the male equivalent. With the rapid advances in 3D bioprinting, it will soon be possible to restore all of the lost body parts back to their original appearance and to return their fully functioning capabilities. Religious, political and cultural leaders, fearful of the loss of power that this will represent to them, will attempt to use every means at their disposal to have laws passed to prohibit such human re-naturalization procedures. But people, just like information, want to be free. They will fight for the right to make their own free choices in life. The technologies described in this book will help give them that freedom.

Trend 28: *Over the very long term, 3D printing will be able to accomplish some truly amazing things on massive scales.*

Implications: Using inter-planetary dust as the primary raw material, it will be possible to manufacture entirely new planets for human habitation and eventually, Dyson Spheres with diameters of 2 AU (Where 1 AU = the earth's average distance from the sun. Such a structure could be used to harvest every photon emitted by a star, generating power levels equivalent to more than 100 trillion times total present human energy consumption.

Overall Conclusions

While business in the 20th century was all about customers, in the 21st century business will increasingly be about *the* customer. The universal availability of 3D printing, scanning and supporting technologies will for the first time make it possible to quickly build inexpensive custom products for each individual buyer.

The power that these technologies will hand over to individuals will be awesome. Anyone, anywhere, with the right vision and with some hard work can harness 3D-making tools to develop an innovative and successful enterprise. They can then use this enterprise to develop and commercialize new products to meet new customer demands as they arise.

It has been shown that the full implementation of all of the various manifestations of Just-On-Order-Making technologies will invariably result in massive societal disruption. A broad swath of established industries will simply be wiped off of the national economic map, and millions of confident employees, who once thought that they were economically secure, at least until retirement, will find themselves without an employer. This is the bad news.

The good news is that while being a job destroyer, 3D printing and supporting technologies will also be a massive job creator. One of the jobs that it will create will be for 3D designers who can take a product idea from a client and then morph it into something that can be brought into real existence by a 3D printer. Designers will specialize in a wide variety of disciplines, including consumer products, medical devices, architectural models, ore body representation for the mining industry and a variety entertainment applications such as the creation of virtual actors. Creative researchers will be required in a variety of areas such as fashion and jewellery design to generate a continuous stream of new consumer product ideas.

Technicians trained in 2D CAD design for the architectural, construction and aero-space industries will be replaced by designers able to work comfortably with 3D models. In addition to 3D designers, industry will need Computer Aided Design/Manufacturing (CAD/CAM) experts able to translate technical ideas into viable operational systems. With the increasingly accurate possibility of testing new vehicle, aircraft or submarine designs in virtual wind/water tunnels, engineers and scientists proficient in the construction of Computation Fluid Mechanics (CFD) and Finite Element (FE) Simulations will be in growing demand.

Biomedical technicians will increasingly be trained to use 3D bioprinters to make medical implants, prosthetics and organs. Chefs trained in the new profession of culinary engineering will use 3D printers to manufacture artistically designed foods.

The new profession of molecular engineering will use atomic-scale 3D printers to assemble, atom-by-atom, new high performance, custom designed materials. Custom designed DNA molecules will lead to the creation of novel, tailor-made living organisms with specifically desirable properties.

Educators at all levels from K – Ph.D. will have to be proficient in a broad spectrum of 3D technologies so that they can train their students to take on the job challenges of the future. All schools will gear up to offer training and certification courses in 3D design software and 3D printing. To attract and retain a new generation of tech savvy youngsters, libraries and community service clubs such as the Scouts, Cubs, Rotary, Lions, Kinsmen and others will be compelled to offer 3D printing activities to their members.

As JOOM technologies become cheaper and more widely available, virtually anyone will be able to make, market and sell products that violate patents held by large manufacturing concerns. Advanced engineering design software will be used to check the designs made by amateurs to assure that all structural components will perform as expected and will not fail under normal conditions of use. While this will decrease the need for structural and design engineers and many related professions, it will at the same time give increased work and opportunities to experts in the areas of IP ownership, patent rights, licensing, fair use, international patent law and other forms of IP protection.

The rapidly falling prices for 3D printers, scanners and supporting technologies means that virtually any small business will be able to enter into the market to design and produce custom made consumer products of all types. These small firms can ramp up quickly and immediately compete will large, long-established industrial concerns. This will result in a highly dynamic, creative, competition-rich environment that will lead to a high level of product innovation. Of course, as far as future employment is concerned, it is somewhat reassuring to realize that all of these new small businesses involved with various aspects of 3D making will still continue to need managers, operations and administrative staff, analysts, engineers, finance and sales professionals, and retail employees, among many others.

Looking into the distant future, it is highly probable that the universal adoption of advanced 3D printers capable of manufacturing anything, anywhere, out of virtual any material, at virtual no cost will usher in a new era of human abundance that may mean that most people will only have to work if they really want to. Certainly this sounds a lot like the dreamy pie-in-the-sky of the the 1950s saying that nuclear generated electricity would be "too cheap to meter." But 3D printers do not need radiation shielding or anti-terrorist fencing and they do not produce highly radioactive wastes that have to be stored for thousands of years. But what they *DO* offer are real environmental, economic, and sociological advantages over conventional manufacturing techniques. For example it requires more than a million gallons of water to manufacture an automobile on a traditional assembly line. A 3D manufactured vehicle requires zero gallons. Like so many numbers associated with the JOOM revolution, these are very hard to argue with.

Welcome to *YOUR* JOOM destiny. Enjoy. Or run like hell. The choice is yours…

Web Resources

3D Printed Car: http://www.youtube.com/watch?v=9kPBtZC8mLc

Future Trends: http://www.youtube.com/watch?v=CD7yB9gZDIk

Future Trends: http://www.youtube.com/watch?v=vk66IIgTEUY

Nano- 3D Printing: http://www.youtube.com/watch?v=2V4DwZmQenY

The Singularity is Near: http://www.youtube.com/watch?v=H4axEZwLdno

Photo Credits

Figure 5-1: Presentation to the US Navy by Dr. Paul Tinari, 2007.

Figure 5-2:
U.S. Navy photo by Photographer's Mate 2nd Class Daniel J. McLain, http://upload.wikimedia.org/wikipedia/commons/2/2a/US_Navy_050627-N-0295M-003_A_full_scale_Joint_Unmanned_Combat_Air_Systems_%28J-UCAS%29_X-45C_on_display_at_the_2005_Naval_Unmanned_Aerial_Vehicle_Air_Demo_held_at_the_Webster_Field_Annex_of_Naval_Air_Station_Patuxent_River.jpg

Figure 5-3:
MC2 Michael Smevog, http://www.defenseimagery.mil/imageRetrieve.action?guid=2abe934f91a632a517466e9c1df5fb0c261de186&t=2

Figure 5-4:
http://www.dfrc.nasa.gov/Gallery/Photo/X-43A/HTML/ED97-43968-1.html

References

Alphonso, C. (2009) "Car, Heal Thyself," *Globe and Mail,* March 13, p. A2

Andreatta, D. (2013) "Science Could Add Decades to the Average Human Lifespan," *Globe & Mail,* Aug. 18.

Bouw, B. (2012) "Miners Look to a Future of Automation," *Globe & Mail,* Feb. 6, p. B3.

Chao, L., Clark, D. (2011) "Cisco Bid Draws Scrutiny," *Globe and Mail,* July 5

Datar, I. (2013) "Test Tube Burgers," *U of T Magazine*, Summer, p. 53.

Doctorow, C. (2009) "*Makers*," Tor Books (US) & Harper Voyager (UK), ISBN 978-0-7653-1279-2

El Akkad, O. (2013) "New Software Allows Insurers to Track Driving Habits and Personalize Premiums," *Globe & Mail*, Aug. 19.

Gardiner, B. (2012) "Robo Blocks," *Popular Science*, February, p. 11

Globe & Mail (2012) "Sky's The Limit for Use of Unmanned Planes," Feb. 13, p.A3.

Gray, J., McFarland, J. (2014) "New Rules Give Boost to Crowdfunding," *Globe & Mail*, March 21, p. B3

Iozzio, C. (2011) "The Cash Killer," *Popular Science*, November, p. 22

Keegan, P. (2009) "The Best New Idea in Business," *Fortune*, September 14, p. 42

Kubota, Y. (2008) "What are You Looking At? Japan Scientists Find Out," *Reuters*, Dec. 17.

Lametti, D. (2010) "Magnets Can Change Your Moral Values," *Discovery*, January-February,

Metalysis (2013) http://www.metalysis.com/titanium

Mitchell, L. (2013) "Clone Wars," *Popular Science Magazine*, January, p. 30

NASA (2013) "NASA's Next Rover Features 3D-Printed Parts," *NASA Tech Briefs*, August, p.20

PS (2012) "Material World," *Popular Science*, pp. 42 - 53

Ramsey, M. (2012) "Ford Invests in Anti-Gridlock Technology,"
Globe & Mail, Report on Business, Feb. 27, p. B3

Ray, S. (2011) "Boeing's 787 Glut Casts $16.2 Billion Cloud Over FAA
Approval," *Bloomberg*, Aug. 23. http://www.bloomberg.com/news/2011-08-
23/boeing-s-787-glut-casts-16-2-billion-cloud-over-faa-approval.html

Scott, A. (2008) "The Bionic Age," *Azure*, March/April pp. 84-87

Shachtman, N. (2008) "The Shield," *Wired*, May, pp. 143 -148

Weber, S. (2004) "*The Success of Open Source*," Cambridge,
MA: Harvard University Press, April.

Wente, M. (2012) "We're Ripe for a Great Disruption
in Education," *Globe & Mail*, Feb. 4

Yglesias, M. (2011) "Reengineering the University,"
Popular Science, September, p.66 – 67.

If You Have Purchased this Book...

You are Eligible for a 20% Discount

On ANY Presentation, Workshop, Seminar or Personal Coaching Session by the Internationally Renowned Futurist, Inventor, Visionary & Speaker,

Dr. Future

*Call Today and Start Inventing **YOUR** Future*

604-760-5088

tinarip@yahoo.com

(Mention Code: JD141)

Dr. Future

Founder JOOM Inc.
Host, *Future Talk* radio show,
Consultant to Fortune 500 Corporations,
Governments around the Globe, NASA and the U.S. Navy

You have read his book, now choose Dr. Future to speak to your organization in person....

With over 25 years if successful speaking experience, Dr. Future's Speaking Generates Positive, Measurable Results!

Dr. Future is a leading authority on the impacts that new, emerging technologies will have on society, governments and on corporate business models.

Need a seasoned professional to brief your decision makers on the most significant trends that will be impacting your business

and your bottom line?

*Do you want to know about these critical trends **before** your competition does?*

Dr. Future has more than 25 years experience as a professional futurist involved in the identification of critical trends before they are readily apparent to government or private sector analysts

Need a distinguished speaker for your next corporate event?

Dr. Future has delivered hundreds of challenging, informative, provocative and entertaining speeches to the most demanding professional audiences

around the globe

Need someone to organize, lead and facilitate an organizational educational seminar?

Dr. Future has been sought out by the most demanding clients to train their top executives and corporate staff.

Contact: Dr. Future Today!

Tel: 604-760-4088   tinarip@yahoo.com

Dr. Future's

Creativity Boot Camp

"Give old, stale ideas and ways of doing things the boot."
"Boot the New Creative You into Existence"

Give your brain a workout like it has never had before...exercise your imagination...learn about new idea-generating techniques and problem solving tools that anyone can use in their workplace to become an amazingly successful team member, leader, manager or supervisor. Stimulate your imagination and gain new perspectives. For starters, you'll learn:

- Little known but highly effective brainstorming techniques you and your teams can use to generate new ideas – you'll be amazed at what you will come up with!

- How to create a work environment that stimulates creativity and innovative problem solving

- Secrets to inventing innovative solutions to problems that would have stumped you in the past

- Innovative ways to transform a group of individuals into a motivated, focused, energized, creative team that consistently produces outstanding results

- How to escape from "thinking black holes" that stifle your natural creativity

- How the very best managers, supervisors and team leaders manage to get their workplaces humming with drive, energy, excitement and enthusiasm.

...and that is only the beginning

Is it possible to have fun and develop an array of incredibly powerful new creative thinking and leadership skills – in one weekend?

You bet! Dr. Future guarantees it.

Contact: Dr. Future Today

Tel: 604-760-5088   E-Mail: tinarip@yahoo.com

Invite Dr. Future to Teach You:

"How to Bring Abundance into Your Life"

The Art & Science of Acquiring Unlimited Personal Wealth

Watching the incredible and bountiful renewal of life each spring is only a tiny demonstration of the powerful creative energy of our universe. The creative universe is a place where any human being with a definite purpose and a burning desire can find and harness available resources to achieve virtually any clearly defined goal.

It is the purpose of this seminar to teach the art and science of generating unlimited personal wealth. The techniques that are taught can be mastered by anyone, regardless of education, talent, previous life experience or previous failure. Abundance is the birthright of every person. This course will teach you how to claim this birthright.

- Some of the practical topics included in this exciting course include:
- Understanding the basic principles of personal wealth generation
- How to gain power over your thoughts and thereby gain control over your fate
- The seven positive emotions and how to manifest them
- The seven negative emotions and how to avoid them
- Developing desire, imagination, decision and persistence while learning to eliminate fear
- Learning how to harness the power of the subconscious mind

And much, much more....

Contact Dr. Future Today

Tel: 604-760-5088

tinarip@yahoo.com

(Mention Code: JD141 and receive a 20% discount)